The Solar System

ZDENĚK KOPAL

The Solar System

OXFORD UNIVERSITY PRESS
London Oxford New York
1973

Library of Congress Catalogue Card Number: 72-96944
This reprint, 1974
Printed in the United States of America

Contents

List of Plates

List of Figures

Acknowledgements

Plate 1 Reproduced by courtesy of the Hale Observatories, California Institute of Technology and the Carnegie Institution of Washington.

Plates 2, 6, 8 Reproduced by courtesy of the Observatoire du Pic-du-Midi.

Plates 3, 4, 10 (*top*) Reproduced by courtesy of G. P. Kuiper.

Plates 5, 9, 11, 12 Reproduced by courtesy of NASA.

Plate 7 Reproduced by courtesy of G. H. Pettengill.

Plate 10 (*bottom*) Reproduced by courtesy of the Jet Propulsion Laboratory, California Institute of Technology.

Plate 13 Reproduced by courtesy of the Yerkes Observatory.

Plate 14 Reproduced by courtesy of D. E. Blackwell.

1
Introduction

MOST OF THE celestial objects that we see in the sky at night are stars. These are objects generically akin to our Sun, and many of them are situated at distances from which light can reach us only after travelling for decades or even centuries through empty space. In fact, the principal characteristic of the space surrounding us in the Universe is its overwhelming emptiness. Of all the stars we can see at night with the naked eye in both hemispheres of our sky, only nine are nearer to us than fifteen light years; and although almost forty more stars are known to lie within this distance, most of them are so faint that large telescopes are necessary to make them visible.

The stars have presented from time immemorial an image of cosmic immutability. This image has been modified in the past few hundred years by the discovery of parallactic and peculiar proper motions of the stars. These demonstrate that the stars in the sky are in fact in perpetual motion, and appear to be immutable only because our measurements of their positions are confined to time spans comparable with human lifetime.

Not all celestial objects visible to the unaided human eye maintain the same relative positions in the sky, like the stars. Since the earliest days of human civilization in Egypt and Mesopotamia five thousand years ago, it has been noted that some of the brightest 'stars' wander in the sky in a complicated manner; and their itineraries have fascinated the human mind ever since. For a long time the number of these objects known remained at five, best known to us by their Latin names: Mercury, Venus, Mars, Jupiter, and Saturn. The first two have been found to be visible alternately in the morning or evening sky; the last three can shine throughout the night.

The apparent motions of these 'wandering stars', or *planets*, were for many centuries interpreted as being caused by their revolutions around the Earth, though very complicated celestial machinery had to be postulated to 'save the phenomena' in this manner. It is true that the idea of a heliocentric solar system had already emerged with Aristarchos of Samos in the third century B.C.; but it flickered briefly during that first 'century of genius', and was rescued from oblivion only by a growing exasperation with its alternatives. In the first part of the sixteenth century—towards the twilight of the Renaissance—Nicolaus Copernicus (1473–1543) revived the heliocentric planetary system, without much success, by dressing it up in the geometrical garb of Ptolemaic epicycles. But it took the discovery of the telescope and a major theological revolution in the first half of the seventeenth century to establish the validity of the heliocentric system as we know it today.

Its laws of motion were laid down by Johannes Kepler (1571–1630) between 1609 and 1618. Kepler deduced these laws from planetary observations made by Tycho Brahe (1546–1601) almost eighty years before Isaac Newton (1642–1727) provided their theoretical justification. Moreover, although the correct model of the planetary system was derived from Tycho's observations by the efforts of Kepler, a realistic approximation to its true scale did not dawn upon an astonished humanity until the end of the seventeenth century.

The astronomical revolution of the seventeenth century, therefore, saw the Earth finally dethroned from its assumed privileged position to become the sixth planet. Moreover, after the advent of telescopic astronomy, the years 1781, 1846, and 1930 brought forth discoveries of three additional planets—Uranus, Neptune, and Pluto—to complete the planetary system as we know it today.

The solar system contains many more objects than just the central star with its nine planets. Several of these planets are attended by families of *satellites* of their own and one, Saturn, is girdled with a ring made up of an untold number of smaller particles. Moreover, the Sun is surrounded by a loose ring of 'minor planets', or *asteroids*, filling up a part of the space between Mars and Jupiter. A large part of the interplanetary space (down to its innermost precincts) is also traversed by solitary eccentric travellers of peculiar structure and composition, generally known as *comets*, whose gradual disintegration (proceeding rapidly on a cosmic time scale) leaves behind them swarms of *meteors*, which become 'shooting stars' if their long cosmic journey happens to be terminated by the Earth's atmosphere.

The aim of this book is to summarize for the non-technical reader the principal features of our present knowledge of the solar system as a whole. In Chapter 2, we shall confine our attention to the major planets—from Jupiter to Neptune—which together constitute the bulk of its mass. These planets have very similar mean densities (close to 1000 kg m^{-3}), which suggests a similarity of chemical composition. Moreover, all major planets occupy a distinct zone around the Sun, the location of which may, in fact, offer a clue to their composition and other more detailed characteristics.

Chapters 3–7 are concerned with another generic group of planetary bodies, of which we can regard our Earth as a prototype. These, too, constitute a group that occupies (with the possible exception of Pluto, the affiliation of which is as yet somewhat uncertain) a specific part of the solar system: namely, its inner precincts. The masses of these planets are more moderate. However, their mean densities are substantially higher (4000–6000 kg m^{-3}), which suggests that their chemical composition is very different from that of the major planets. In point of fact, while the major planets consist of material that is essentially solar in composition (i.e. hydrogen and helium), the terrestrial planets, of which our Earth is the most massive, are made up of heavier elements (predominantly oxygen, silicon, and iron). The structure and composition of the Earth have already been treated extensively in many other books available to the reader, so we shall confine our attention mainly to the astronomical aspects of our mother planet and its comparative place in the solar system.

Chapter 7 is reserved for the Moon. By its mass, size, and composition the Moon belongs to the tail of the terrestrial planets. It has recently become the target of man's entry into the wider arena of the solar system and has yielded so many of its surprising secrets to direct exploration, both manned and unmanned, that a separate chapter is needed to do justice to this newly acquired knowledge.

The comets, floating icebergs of frozen hydrocarbons traversing space, are the subject of Chapter 9. Chapter 10 is concerned with cosmic material left in the wake of comets and other debris encountered in interplanetary space.

Chapter 11 gives a retrospective view of the salient features of the evidence assembled in Chapters 2–10, in an attempt to outline what is known of the origin and evolution of the solar system as a whole. Chapter 12 raises the question: 'Are there other planetary systems in the Universe around us, or to what extent can our own solar system be considered as unique?' Although we are still far from being able to reconstruct the details of the process by which our solar system originated, observational evidence already available suggests that other stars in our neighbourhood are attended by planets comparable in mass to Jupiter. What is more, such planets, far from being rare in the Universe, may actually be very common. Chapter 13 gives a brief outline of the likely trends of future planetary research.

The foregoing list omits the central body of this system—our Sun. It might seem that to write a book about the solar system without the Sun is like playing Hamlet without the Prince of Denmark; for the Sun is overwhelmingly more massive than all the rest of the bodies attending it, representing as it does more than 99·87 per cent of the total mass of the entire solar system. A disparity in mass so overwhelming is, however, bound to render the Sun a physical structure utterly different from the planets. The only property the two classes of celestial bodies may have in common is their initial chemical composition. Moreover, the physics of the Sun is well covered by many other existing sources. In what follows we shall, therefore, regard the Sun merely as fulfilling a twofold purpose: it provides a gravitational anchor for all the bodies revolving around it; and it is a source of light illuminating the planetary surfaces, the absorption of which maintains them at their observed temperatures. Radiation from the Sun has, indeed, provided the energy for virtually all the processes that have shaped the planetary surfaces; only nuclear power derived from the slow spontaneous disintegration of radioactive potassium, or certain heavy elements like uranium or thorium, represents an independent source whose origin may antedate the Sun. All other processes, and, in particular, the erosion caused by circulation of the atmospheres or oceans, have been maintained by sunlight—contemporary or fossil. The influence of the Sun constantly pervades every part of its entire system.

Within the self-imposed limits of the present book, our aim will be to outline for the reader the present state of our knowledge of the astronomy, physics, and chemistry of the solar system, with emphasis on the most recent work. Limitations of space alone would prevent us from tracing the development of our knowledge in any kind of historical perspective. There is, in fact, ample justification for emphasizing recent work: most of the knowledge we now possess has been acquired only since methods of classical astronomical approach have been supplemented—and largely superseded—by more modern methods of radio, radar, and space astronomy based on the use of spacecraft.

These newer methods have truly revolutionized our subject in recent years, and their impact on it is still far from being exhausted. Their contributions have already, however, been on so massive a scale that all books written about the solar system before the advent of the space age could as well have been written in Latin or Greek, so dated do they appear to a contemporary reader.

Moreover, as often happens at particular crossroads in science, these striking advances in our knowledge of the solar system could have been accomplished only by a concerted interdisciplinary effort in which investigators from many different branches of science took part. In fact, exploration of the solar system is no longer a domain of professional interest only to astronomers. Radiophysicists, chemists, geologists, engineers, and astronauts have joined forces to bring about advances that would have made our astronomical ancestors gasp in awe.

2
The Major Planets of the Solar System and their Satellites

THE LARGEST PART of the mass of our solar system (other than that constituting its central star) and almost all of its angular momentum is stored in four planets—Jupiter, Saturn, Uranus, and Neptune. These revolve around the Sun at distances ranging from 778 million km for Jupiter to 4500 million km for Neptune, in orbits of very small eccentricity* situated almost in the same plane (Figure 1). If the distances are expressed in terms of that separating the Earth from the Sun (which in solar-system astronomy we regard as our astronomical unit*) Jupiter is

TABLE 1
Kinematic properties of the major planets

	Jupiter	*Saturn*	*Uranus*	*Neptune*
Mean distance A from the Sun (in AU)	5·204	9·548	19·20	30·08
Light transit time (in minutes)	43·27	79·34	159·5	250·1
Orbital period P (in years)	11·86	29·46	84·02	164·79
Orbital eccentricity	0·048	0·056	0·047	0·009
Orbital inclination to the ecliptic	1° 18′·4	2° 29′·5	0° 46′·3	1° 46′·6
Period of axial rotation (on the equator) (i.e. length of day)	$9^h\ 50{\cdot}5^m$	$10^h\ 14^m$	$10^h\ 45^m$	$15{\cdot}8^h$
Inclination of equator to orbital plane	3°·1	26°·7	98°·0	29°·0

found (see Table 1) to revolve round the Sun at a mean distance of 5·2 astronomical units (AU) and Neptune at 30·1 AU. Light from the Sun (which needs just under 500 seconds or a little over eight minutes to reach

* Words marked in this way are described either in a footnote or in the Glossary at the end of the book.

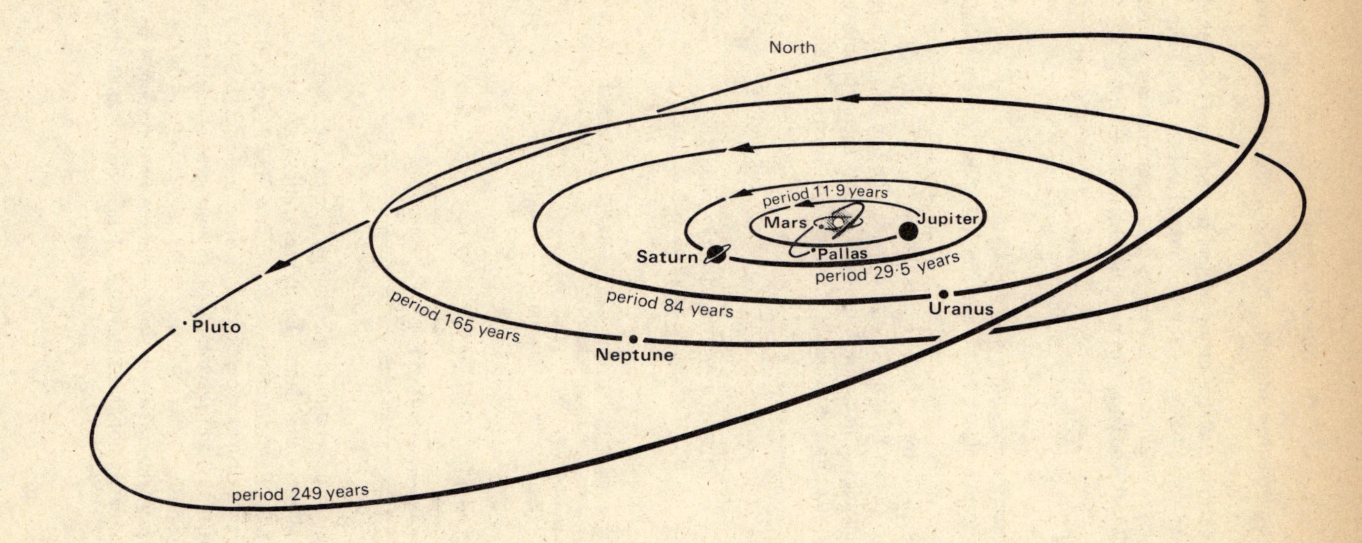

FIG. 1: The orbits of the outer planets of the solar system.

the Earth) will traverse these distances in times ranging between 43 and 250 minutes.

The orbital periods P of the major planets range from 11·86 terrestrial years for Jupiter to 164·8 years for Neptune (Table 1); and their mean angular velocities $\omega = 2\pi/P$ are related to their mean distance A from the Sun by Kepler's well-known law of periodic times, of the form

$$\omega^2 = \frac{G(m_\odot + m)}{A^3},$$

where $m_\odot$ denotes the absolute mass of the Sun ($1{\cdot}986 \times 10^{30}$ kg); m the mass of the respective planet, and G the absolute value of the gravitational constant ($6{\cdot}668 \pm 0{\cdot}005 \times 10^{-14}\ \mathrm{m^3\,g^{-1}\,s^{-2}}$).

From values of ω (or P) and A established by observations of each respective planet (and a known value of $Gm_\odot$ for the Sun), Kepler's law of periodic times enables us to establish the mass-ratio of each planet to that of the Sun ($m/m_\odot$). Reciprocal values of this ratio are listed in Table 2, with

TABLE 2
Physical properties of major planets

	Jupiter	*Saturn*	*Uranus*	*Neptune*
Mass-ratio $m_\odot/m$ (Sun : planet)	1047·36	3498·7	22 692	19 349
	±·01	±·2	±33	±28
Mass-ratio $m/m_\oplus$ (planet : Earth)	317·89	95·16	14·67	17·21
Equatorial diameter (km)	142 700	120 800	47 300	50 600
	(11·20⊕)	(9·47⊕)	(3·71⊕)	(3·97⊕)
Mean density ρ_m (kg m^{-3})	134	69	156	152
Ratio ρ_c/ρ_m	3·1	6	1·8	2·1
Surface temperature (noontime) in K	135°	105°	65°	50°

indications of their observational uncertainty. The second row contains the corresponding value of the planet : Earth mass-ratios (computed from the fact that the Sun : Earth mass-ratio, $m_\odot/m_\oplus$, is $332\,945{\cdot}6 \pm 0{\cdot}3$). The masses of the major planets range, therefore, from almost 318 ⊙ for Jupiter to 14·7 ⊙ for Uranus; and the total is 0·134 per cent of the mass of the Sun.

Thus, the overwhelming part of the *mass* of the solar system is stored in the Sun. The same is not true, however, of the second most important physical attribute, the *angular momentum*. In point of fact, while some 99·87 per cent of the mass of the solar system resides in the Sun, an almost equally large preponderance (98·1 per cent) of its total momentum is represented by the orbital momenta of the major planets. This disparity is of great significance for the origin of the system as a whole, and we shall return to it again in Chapter 11.

The orbital momenta of the major planets do not, however, represent the total angular momenta of these bodies; for all major planets *rotate* with relatively high angular velocities (the periods are listed in Table 1). For all major planets but Uranus the axes of rotation are only moderately inclined to the orbital plane, and the sense of rotation is direct.* The 'day' on the major planets is generally short in comparison with the terrestrial one: less than half of our own day for all but Neptune, and only 9 hours and $50\frac{1}{2}$ minutes for Jupiter. This latter value only refers, however, to the duration of the Jovian sidereal day* on the *equator*. At higher latitudes, the rotation is slower by more than 5 minutes. On Saturn, the length of the day on the equator is equal to 10 hours 14 minutes; at $\pm 40^{\circ}$ latitude it has increased by 27 minutes to 10 hours 41 minutes. These facts disclose that neither Jupiter nor Saturn rotates like a rigid body; instead, their globes behave as fluid bodies capable of differential rotation. This, in turn, provides an important indication of their internal structure—a problem to which we shall presently turn our attention.

Before we do so, however, we must detail one additional important characteristic of the major planets: their size. We can, in principle, measure the apparent diameter of a planetary disc at a known distance and convert this to absolute units by triangulation. The apparent diameters of the major planets are easily resolvable through our telescopes: Jupiter with its 50″ maximum angular diameter at the time of nearest approach of 4·1 AU can almost be seen as a finite disc with the naked eye (and easily so through binoculars). Saturn, more distant, never shows a disc larger than 17″·3 across, which requires a small telescope for resolution. The angular diameters of Uranus and Neptune amount to only 3″·8 and 2″·3; these planets can be seen as measurable discs only through moderately large telescopes. As the absolute errors of angular measurements are approximately the same regardless of absolute dimensions, the proportional errors inherent in our determinations of the absolute dimensions of Uranus and, especially, Neptune are much larger than those for Jupiter and Saturn. In point of fact, the absolute dimensions of Neptune were only recently properly determined from the duration of the occultation* by Neptune of a star (BD -17° 4388) on 7 April 1968.

The diameters of the major planets are given in Table 2, both in kilometres and terrestrial units. These are the equatorial diameters of the respective planets; the centrifugal force due to axial rotation flattens their globes to an appreciable extent, and renders their polar diameters almost 10 per cent shorter than the equatorial ones.

Jupiter, the largest planet of the solar system, is more than eleven times larger than the Earth, and less than ten times smaller than the Sun in linear dimensions; Saturn, the second-largest planet, is 9·5 times larger than the Earth; but Uranus and Neptune are only 3·7 to 4·0 times as large. However, when we divide the masses of these planets by their respective volumes, we find that their mean densities are very much alike. For Jupiter, Uranus, and Neptune they are close to 1300 to 1500 kg m^{-3} (Table 2); only Saturn, with its mean density of less than 700 kg m^{-3}, appears to be somewhat anomalous.

Measurement of the apparent angular diameter of a major planet, combined with its observed apparent brightness, enables us to determine the reflectivity of its surface. At the time of its opposition* (i.e. minimum distance of 4·1 AU from us), Jupiter shines in the sky as a star of −2·5th apparent visual magnitude*—the brightest planet in the sky after Venus. This brightness indicates that approximately 44 per cent of the incident sunlight is scattered by the Jovian surface. The maximum apparent brightness of Saturn (−0·4th magnitude) indicates that this planet scatters only 42 per cent of the incident sunlight. Uranus, with an apparent brightness of $5^m{\cdot}5$, is just visible to the naked eye at the time of closest approach (18 AU), and has a reflectivity expressed by an albedo* of 0·45. Neptune with a similar albedo appears (at 29 AU) as a star of visual magnitude only 7·8; it requires a small telescope to make it visible at all.

Internal constitution of the major planets

Two clues to internal constitution are already in our hands: the mean density of the planetary globes and the polar flattening due to axial rotation —i.e. the extent to which the sphere as a whole yields to the centrifugal force. The amount of this flattening leads to values for the ratio of the central density to the mean density of the respective planets (ρ_c/ρ_m). These values, given in Table 2, disclose only a moderate degree of central condensation—larger than that obtaining in our Earth (in which the ratio ρ_c/ρ_m is equal to about 3), but much smaller than in the Sun or stars.

On the other hand, any cosmic mass as large as that possessed by a major planet of the solar system is bound to be in hydrostatic equilibrium, regardless of the form of the equation of state* which holds in its interior. The internal pressure should, therefore, be largely determined by the distribution of mass alone. Accordingly, the pressure at the centre of Jupiter would be expected to exceed 10^{13} N m^{-2}; and in Saturn, to be about half this value. Such pressure should be sufficient to compress the constituent material of these planets to central densities consistent with the values of ρ_c/ρ_m inferred from the observed extent of polar flattening.

The mean densities ρ_m disclose that the bulk of the mass of all the major planets must consist, like the Sun and the stars, of the lightest two elements, hydrogen and helium; for no other mixture could render them so light. Indeed, recent computations (1970) of models indicate that Jupiter may consist of approximately two-thirds of hydrogen and one-third of helium by weight (i.e. a somewhat lower proportion of hydrogen to helium than in the Sun). Saturn may possess an even smaller proportion of hydrogen, in spite of its lower mean density; for its smaller mass will subject it to less compression. The balance of the mass in Saturn, as in Jupiter, is undoubtedly helium; though in Uranus and Neptune some carbon, nitrogen, and oxygen may be present to account for their mean densities, which are of the order of 1500 kg m^{-3} in spite of the smaller degree of compression caused by their mass.

What temperatures are likely to be attained in the interiors of the major planets? In order to account for their densities, it is necessary to assume that the bulk of hydrogen must be present in their interiors in the *solid* state; and this requirement by itself imposes an upper limit on their temperatures. Under the pressures prevailing in the Jovian or Saturnian interiors, hydrogen can remain in the solid (metallic) state up to temperatures near 7500 K, but no higher; this is probably the limit (about the same as we shall find inside the Earth) which their internal temperatures cannot exceed. A similar argument applies to helium which, in contrast to hydrogen, does not solidify at any pressure inside the planets. If the relative abundance of helium inside Jupiter and Saturn were greater than was allowed for in our foregoing estimates, gaseous helium would become very much more compressed than solid hydrogen, and the central condensation of these planets would be higher than is consistent with their observed flattening at the poles. The presence of a metallic core, at least in Jupiter, is confirmed by a magnetic field emanating from its interior, as attested by the polarization of its emission at radio-frequencies. Whether or not Saturn, Uranus, and Neptune possess similar fields has not yet been settled by observation, but it remains a distinct possibility.

It is only natural to expect that celestial bodies of the mass and composition of the major planets should have at their surfaces extensive semi-

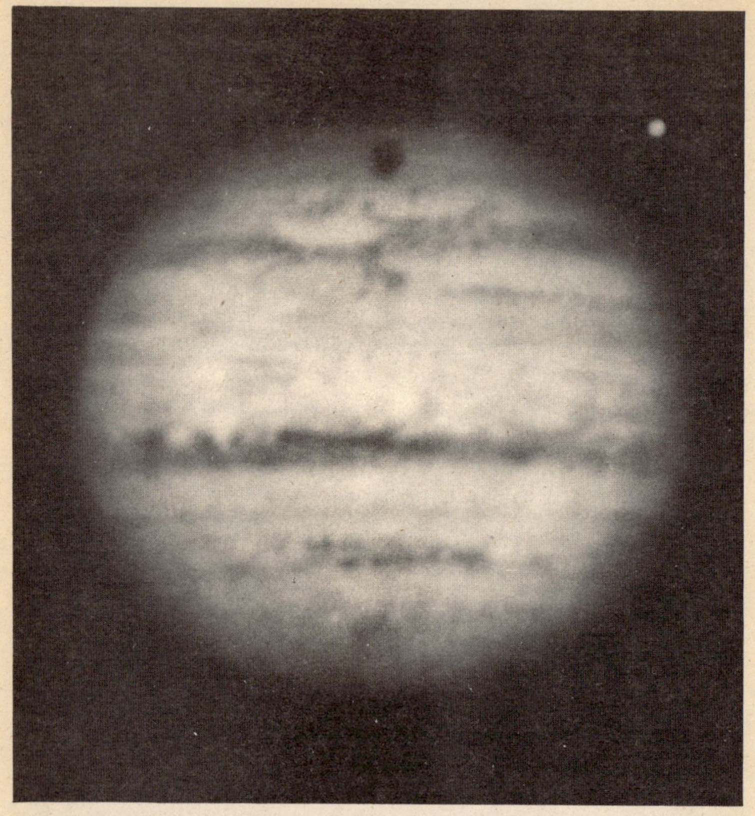

PLATE 1: Photograph of the planet Jupiter, in red light, taken with the 200-inch reflector at Mount Palomar. The satellite Ganymede (Jupiter III) can be seen in the upper right corner, with its shadow cast on the planetary disc.

transparent layers, which for want of a better word we can call their *atmospheres*. That this is indeed so has been disclosed by observations almost since the time of the discovery of the telescope, and all the features we can see or photograph on the apparent discs of Jupiter and Saturn (see Plates 1 and 2) are essentially atmospheric. This is particularly true of the dark bands on Jupiter (Plate 1) aligned with the planetary equator by centrifugal force due to axial rotation; or of the famous red spot, the position of which continues to drift slowly in the course of time.

The spectra of all major planets disclose evidence of powerful absorption

PLATE 2: Photograph of the planet Saturn and its rings, taken by P. Guérin with the 43-inch reflector of the Observatoire du Pic-du-Midi.

by molecular hydrogen and methane, followed by ammonia (whose absorption bands are strong in the spectrum of Jupiter and weak in that of Saturn), together with traces of the absorption by acetylene, hydrogen sulphide, and hydrogen cyanide. Helium is no doubt present in large amounts, but it is impossible to detect its spectroscopic effects in the ultraviolet through the ozone layer of our own air.

The temperatures prevailing in the semi-transparent outer layers of the major planets are generally low. If the radiation received from the Sun were the only source of heat available to these planets, their surface temperatures should be close to 105 K at the distance of Jupiter, and 78 K for Saturn; at Uranus and Neptune temperatures as low as 55 K and 43 K should be attained. The actual temperatures of these planets, as deduced from the intensity of their thermal radiation in the near infrared, are given in Table 2. The reader may note that the actual temperatures of Jupiter and Saturn are twenty to thirty degrees higher than those of a black body exposed to sunlight at the respective distance, and correspond to a flux of radiation almost twice as large as that which could be explained by sunlight alone. This fact suggests that the major planets—in particular, Jupiter and Saturn—must possess small stores of internal heat which supplement that received from the Sun. The source of this heat is probably secular* contraction of the respective planetary globes, and it can be shown that a diminution of their radii by as little as 1 mm per year should be sufficient to account for all that is needed.

Radio noise

Radiation emitted by the major planets in the centimetre domain of their radio spectra indicates, in general, somewhat higher temperatures than those inferred from the infrared flux (no doubt owing to the fact that, at longer wavelengths, we penetrate deeper into the atmosphere where higher temperatures prevail). For decimetre waves the corresponding temperatures are very much higher. However, the radio spectrum of Jupiter at decametre wavelengths (wavelengths greater than 7·5 m) held the real surprise: the continuous spectrum with constant intensity exhibits transient outbursts so intense at times that their origin must be non-thermal. Such explosions of Jovian noise last only milliseconds, and may be associated with individual thunderbolts of electrical storms in the atmosphere which rage much longer and are strongly localized. The radiation emitted by them seems to be largely directional, and intensive outbursts recur whenever the trouble-spot happens to pass through the meridian* for the terrestrial observer. One such particularly conspicuous trouble-spot happens to pass through the meridian every 9 hours 55 minutes 29·73 seconds, with an uncertainty of ±0·04 second. Since the angular velocity of axial rotation on Jupiter varies between the equator and the pole, the observed value permits us to localize the latitude of this permanent disturbance at ±49°.

The radiation of the bursts at decametre wavelengths is, moreover, linearly polarized. This can scarcely be accounted for otherwise than by synchrotron radiation of electrons gyrating in a strong magnetic field. The strength of this field has been estimated at several hundred gauss at the source (i.e. about a thousand times as great as the strength of the magnetic field of our Earth), though whether a field of such strength is general or merely local remains as yet uncertain. It seems to be established, however, that Jupiter is surrounded by extensive lobes of van Allen belts of charged particles trapped in its magnetic field, whose agitation leaves a distinct imprint on the radio spectrum of the planet. It can also be confidently predicted that Jupiter exhibits extensive auroral activity around its magnetic poles, of which some indications have in recent years been reported on photographs taken in monochromatic light.

Another interesting feature of the radio noise emanating from the Jovian environment is the fact that its decametric component appears to fluctuate in intensity, by as much as a factor of five, with the relative position of Jupiter's satellite Io—the innermost of its four large Galilean satellites—in the sense that the intensity of emission becomes a maximum whenever Io happens to be at quadrature* with respect to the Earth. The reason for this curious phenomenon is probably the fact that Io revolves within the Jovian radiation belts, and disturbs their material in such a way that the radio noise becomes a maximum whenever the wake behind the revolving satellite is parallel with the line of sight.

Satellites

None of the four major planets of our system travels around the Sun alone. They are attended by small families of satellites: two are known so far for Neptune, five for Uranus, nine for Saturn*, and twelve for Jupiter (Plates 3 and 4). Each of these planets controls its satellites gravitationally in much the same way as the Sun controls its planets. The satellite systems can, in fact, be regarded as miniature editions of the solar system as a whole, and may indeed have originated in a similar way.

Of these twenty-eight satellites (plus three more that we shall encounter in the next chapter), only six are comparable in mass and size to our Moon.

* The discovery of a tenth satellite of Saturn, reported by Dollfus in 1966, as yet lacks independent confirmation.

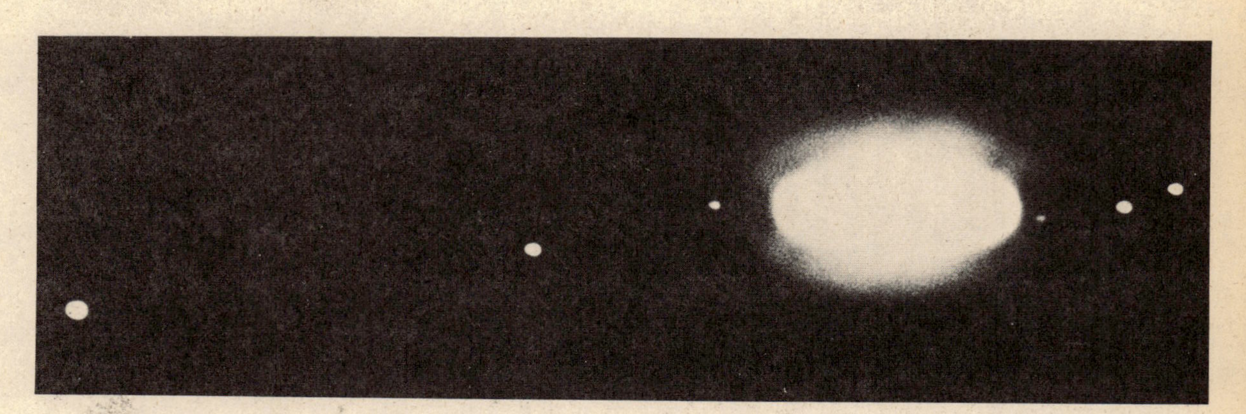

PLATE 3: Six inner satellites of Saturn, photographed by G. P. Kuiper with the 82-inch telescope of McDonald Observatory.

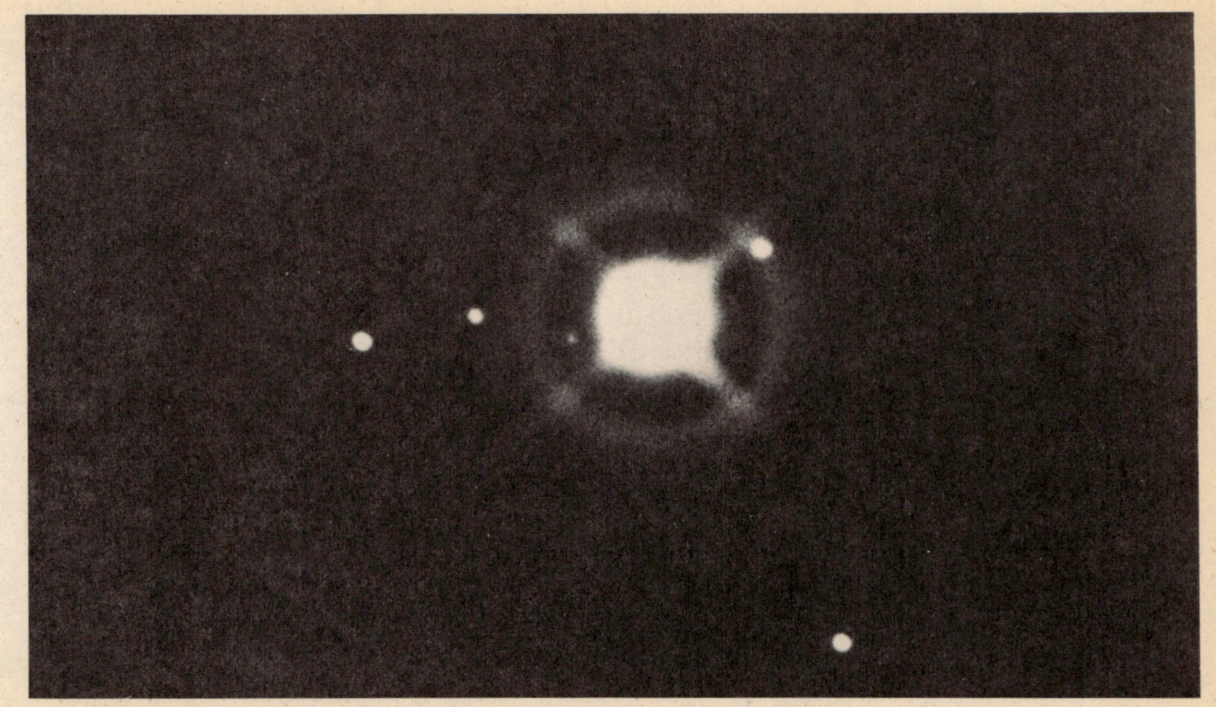

PLATE 4: The planet Uranus with its five satellites, photographed by G. P. Kuiper with the 82-inch telescope of McDonald Observatory.

These are the four Galilean satellites of Jupiter: Io, Europa, Ganymede, and Callisto (discovered by Galileo Galilei and Simon Marius in 1610), one satellite of Saturn (Titan), and one of Neptune (Triton). The ratios of their masses to that of their central planet are of the order of 1 : 1000 or less—i.e. of the order of that which the mass of Jupiter or Saturn bears to the Sun, rather than our Moon to the Earth (see Table 3).

Most satellites of major planets revolve around their central bodies in orbits of small eccentricity, only slightly inclined to the planet's equator (which means that, for Uranus, the satellite orbits are nearly perpendicular to the orbital plane of the planet) and in the direction of the planetary rotation. Exceptions to this are, however, known to exist. Thus, Neptune's satellite Triton revolves around its central planet in a retrograde direction (i.e. opposite to that of the planet's axial rotation); and the same is true of Saturn IX (Phoebe) and Jupiter satellites VIII–IX and XI–XII. The last four (and Phoebe) may, in fact, be simply captured asteroids or comets; the directions of their orbital motions could then be a consequence of initial conditions at the time of the capture; but Triton's retrograde motion is more difficult to account for.

The orbital periods of the satellites range from 0·5 to 10·7 days for Jupiter I–V, 0·9 to 21·3 days for Saturn I–VII, and 1·4 to 13·5 days for Uranus I–V; Jupiter VI–XII, Saturn VIII–IX, and Neptune II take from several months to almost two years to circumnavigate their respective planets. Whether the latter represent true-born children of their central bodies, or mere foundlings only temporarily attached to the planetary girdle, remains as yet uncertain. A possibility that the latter is the case is indicated by the fact that—in their distances from Jupiter—satellites (I–V), (VI–VII and X), and (VIII–IX and XI–XII) form almost three distinct groups revolving around the planet in relatively narrow domains within 1·8 million, around 12 million, and between 21 and 24 million km from the planet, with wide empty gaps in between.

Two of the Galilean satellites of Jupiter (Ganymede and Callisto) and Saturn VI (Titan) are not merely comparable in size with the Moon, but are actually larger than Mercury by a small margin. Their mean densities, however, range only between 1500 and 3000 kg m^{-3} and are, therefore, distinctly lower than that of our Moon and less than half of that of Mercury. Their apparent discs are generally too small for us to be able to discern reliably many surface details. However, from (small) variations in the light they reflect as they revolve, we surmise that their surfaces are not uniformly bright, and that their axial rotation has been synchronized with revolution around Jupiter by powerful tidal action of the central planet. None of the

Table 3
Satellites of the major planets

Satellite (name)	*Orbital period (days)*	*Fractional orbital dimensions A/R*†	*Mass-ratio satellite : planet*	*Radius (km)*
	Jupiter ($R = 11{\cdot}20\oplus$, $m = 317{\cdot}89\oplus$)			
V (Amalthea)	0·498	2·539	10^{-9}	70
I (Io)	1·769	5·905	$3{\cdot}8 \times 10^{-5}$	1670
II (Europa)	3·551	9·396	$2{\cdot}5 \times 10^{-5}$	1460
III (Ganymedes)	7·155	14·99	$8{\cdot}2 \times 10^{-5}$	2550
IV (Callisto)	16·69	26·36	$5{\cdot}1 \times 10^{-5}$	2360
VI	150·6	160·7		50
VII	260·1	164·4		10
VIII	260	164		8
IX	617	290		6
X	692	313		10
XI	735	326		7
XII	758	332		8
	Saturn ($R = 9{\cdot}47\oplus$, $m = 95{\cdot}16\oplus$)			
I (Mimas)	0·942	3·111	$6{\cdot}7 \times 10^{-8}$	280
II (Enceladus)	1·370	3·991	$1{\cdot}3 \times 10^{-7}$	320
III (Tethys)	1·888	4·939	$1{\cdot}1 \times 10^{-6}$	500
IV (Dione)	2·737	6·327	$1{\cdot}8 \times 10^{-6}$	480
V (Rhea)	4·518	8·835	$4{\cdot}0 \times 10^{-6}$	670
VI (Titan)	15·95	20·48	$2{\cdot}4 \times 10^{-4}$	2440
VII (Hyperion)	21·28	24·83	$2{\cdot}0 \times 10^{-7}$	300
VIII (Japetus)	79·33	59·67	$2{\cdot}0 \times 10^{-6}$	550
IX (Phoebe)	550·4	216·8		110
	Uranus ($R = 3{\cdot}71\oplus$, $m = 14{\cdot}67\oplus$)			
I (Miranda)	1·413	5·494	$1{\cdot}0 \times 10^{-6}$	90
II (Ariel)	2·520	8·079	$1{\cdot}4 \times 10^{-5}$	280
III (Umbriel)	4·144	11·25	$5{\cdot}9 \times 10^{-6}$	250
IV (Titania)	8·706	18·46	$5{\cdot}0 \times 10^{-5}$	450
V (Oberon)	13·46	24·69	$2{\cdot}9 \times 10^{-5}$	450
	Neptune ($R = 3{\cdot}97\oplus$, $m = 17{\cdot}21\oplus$)			
I (Triton)	5·877	15·85	$1{\cdot}3 \times 10^{-3}$	2100
II (Nereid)	359·4	249·5	$3{\cdot}0 \times 10^{-7}$	100

† The ratio of the mean radius of the satellite's orbit round its planet to the radius of the planet.

large Jovian satellites exhibits any indication of the presence of an atmosphere. The only satellite in the solar system that does so is Saturn's Titan, whose spectrum shows traces of the absorption bands of methane; the light from the rest remains a faithful replica of the illuminating sunlight.

Because of their proximity to the central planet and the location of their orbits in the planet's equatorial plane, the Galilean satellites of Jupiter frequently enter the shadow of the planet to undergo occultations, as well as to exhibit transits across its disc (thus causing total eclipses of the Sun on the Jovian surface; see Plate 1). Before the satellite disappears completely in the planet's shadow (or after it emerges), the light reflected from it will be attenuated by passing through atmospheric layers of increasing density; spectral observations of this attenuation provide valuable insight into the stratification of different atmospheric constituents with altitude.

All other lesser satellites in the systems of Jupiter and Saturn are too small for us to determine their size directly. By measuring their apparent brightness and assuming a certain surface reflectivity we can, however, at least estimate their size. Their diameters prove to be generally less than 1000 km, and in many cases only a few hundred. From the relatively large variations of light reflected during the course of their revolution we infer that many (in particular Saturn VIII, Japetus) have shapes which are irregular rather than spherical. Similar light variations are common among the asteroids of the solar system (Chapter 8), which revolve around the Sun in great numbers in the space gap between Mars and Jupiter. This is further evidence that at least some of the outer Jovian satellites (in particular, Jupiter VIII–IX or XI–XII) are nothing else but captured asteroids.

Saturn's rings

The concentric system of rings surrounding Saturn (Plate 2) is situated exactly in its equatorial plane, and extend outwards to an abrupt edge 139 000 km from the planet (2·28 times the planetary equatorial radius). Its inner boundary peters out very much more gradually, and according to investigations with electronic cameras (Rösch *et al.*, 1970) it extends like a curtain all the way down to the planet's surface.

When seen edgewise (twice in the course of Saturn's revolution around the Sun, at intervals separated by 13·72 and 15·74 years) the rings disappear completely from sight for several days. This demonstrates that they are extremely thin: not more than kilometres, or even metres, of vertical extension. A star occulted by the rings—a rare event, but one observed repeatedly in the past—remains visible through them without any apparent diminution of its light. Such phenomena demonstrate that the rings do not consist of material spread out continuously in their plane, but rather of an immense number of small particles (akin to the meteors or meteorites). Each of these revolves around Saturn independently of all the others, with a Keplerian angular velocity appropriate for its distance from the planet's centre.

The total mass of the ring (i.e. the sum of the masses of all the individual particles constituting it) appears to be immeasurably small—certainly less than one-millionth of the mass of Saturn itself. Some parts of the rings appear to be brighter (i.e. scatter more sunlight) than others. This can be explained by a greater density (or different size) of individual particles, per unit volume, at the respective location. At certain distances, there appear to be virtual gaps in the rings, of which the Cassini division, at 1·95 times the planet's radius from its centre, is the most conspicuous. These are probably explained by the instability of the orbits of individual particles of a particular size, owing to dynamical perturbations by the satellites (in particular, Titan). Any particle that happened to drift into such a region would soon be ejected by perturbations; relatively few will therefore be seen there at any particular time.

The most likely theory for the origin of these rings is that a former satellite ventured too close to its central planet and was torn to pieces by its attraction. That such a process is dynamically possible was proved a long time ago, but whether this event actually occurred in the Saturn system is not yet known.

Whatever its origin, a dense ring of small particles does not represent a configuration that can remain stable over intervals of time comparable with the age of the solar system. We do not know when Saturn acquired its present ring, or if this is the first ring it had. It is possible that Jupiter likewise once possessed a ring, or may acquire another if Amalthea (Jupiter V satellite) wanders too close to its surface, though none of us is likely to live long enough to see this happen.

3
The Earth

IN THE PREVIOUS chapter we discussed the major planets of our solar system and their satellites. The masses of these planets are a thousand to ten thousand times smaller than that of the Sun, but their chemical composition is very much akin to that of our central star. When, however, we turn our attention to the inner precincts of our solar system (Figure 2), we find it populated by an entirely different group of planetary bodies, which we call the terrestrial planets. These differ from the major planets located outside in

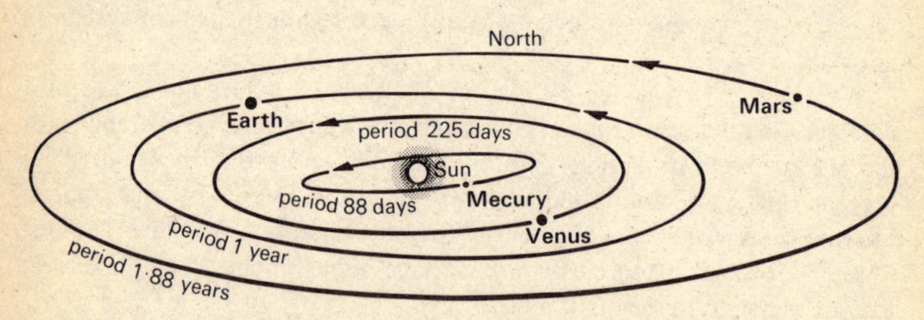

FIG. 2: The orbits of the terrestrial planets of the solar system.

chemical composition even more profoundly than they do in mass or size. The principal kinematic and physical properties of these planets are summarized in Tables 4 and 5.

It is natural for us to turn our attention among them first to the Earth; for a knowledge of its structure, composition, and other physical properties will provide a natural basis for an understanding of the other terrestrial planets which are discussed in Chapters 4–7.

By its physical characteristics, our Earth cannot claim any particular distinction among the members of our planetary family—let alone in the Universe at large. However, in one respect it is unique, at least in the confines of our solar system; for we now know it is the only planet that, in the course of its cosmic evolution, gave rise to life.

TABLE 4
Kinematic properties of terrestrial planets

	Earth	*Venus*	*Mars*	*Mercury*	*Pluto*
Mean distance from the Sun (AU)	1	0·7233	1·5237	0·3871	39·52
Light transit time (minutes)	8·32	6·02	12·67	3·22	328
Orbital period	$365^d.256$	$224^d.701$	$686^d.980$	$87^d.969$	$249^y.17$
Orbital eccentricity	0·0167	0·0068	0·0933	0·2056	0·249
Orbital inclination to the ecliptic*	0	$3° 23'.6$	$1° 51'.0$	$7° 0'.2$	$17° 6'$
Duration of sidereal day	$23^h56^m4^s$	$-243^d.1$	$24^h37^m23^s$	$58^d.6$	$6^d.39$
Inclination of the planet's equator to its orbital plane	$23°.5$	$2°.2$	$25°.2$	0	?

TABLE 5
Physical properties of the terrestrial planets

	Earth	*Venus*	*Mars*	*Mercury*	*Pluto*	*Moon*
Mass (Earth = 1)	1	0·8149	0·1076	0·053	0·18	0·0123
Mean radius (km)	6371	6056	3394	2440	3500	1738
Mean density (kg m^{-3})	5530	5230	3940	5500	~6000	3340
Atmospheric pressure (bar)	1·013	>100	0·0065	none	?	none
Principal atmospheric constituents	N_2, O_2	CO_2, N_2	CO_2, N_2, or Ne?	none	H_2?	none

Our ancestors regarded the Earth as a flat disc, supporting on its edge the sky overhead, and covering up the underworld. The idea that the Earth was a sphere emerged in the human mind some time in the fifth or sixth century B.C. with the Pythagoreans, though the reasons on which this concept was based were probably at first mainly aesthetic. The first scientific argument supporting the theory that the Earth is a sphere appears to have been advanced in the fourth century B.C. by Aristotle, who stated in his book *De*

Coelo that this must be so because the outline of its shadow cast on the Moon during lunar eclipses was always circular.

Size and shape

Having decided that the Earth was a sphere, the early scientist philosophers were naturally curious to know its size. It was obvious to Aristotle and his contemporaries that the Mediterranean world in which they lived must constitute but a very small part of the total terrestrial surface. In the following century, Eratosthenes of Cyrene—a great geographer who spent most of his life in Alexandria—succeeded in determining the dimensions of the Earth by triangulation of an arc of meridian between Syene (Aswan) and Alexandria. He was correct to within a few per cent, so that a knowledge of the shape and size of our planet has been with mankind for some 2300 years. In more modern times, the size of the Earth has been re-measured by techniques essentially similar to those initiated by Eratosthenes. The equatorial radius is 6378·20 km; the polar radius being 21·40 km shorter. The Earth is, therefore, not exactly a sphere, but a rotational spheroid flattened at the poles by centrifugal force arising from diurnal rotation—a fact first noted in the latter part of the seventeenth century.

Rotation and revolution

The Earth revolves around the Sun in one sidereal year of 365·256 36 mean solar days (or 365 days, 6 hours, 9 minutes, 9·348 seconds), along an elliptical orbit of small eccentricity ($e = 0{\cdot}016\,72$). The semi-major axis* of this orbit (equal to our mean distance from the Sun) is now known to be very close to 149 597 900 km—a distance commonly referred to as the 'astronomical unit' (AU) of planetary astronomy. The plane in which the Earth revolves around the Sun is generally known as the ecliptic. Its outline is marked in the sky by the apparent motion of the Sun.

The yearly revolution around the Sun is not the only motion performed in space by the Earth. It also rotates in the period of 23^h56^m and $4^s{\cdot}0906$ of mean solar time about an axis inclined to its orbital plane, in such a way that the angle between the terrestrial equator and the ecliptic amounts (at present) to $23°27'17''{\cdot}6$ (the 'obliquity of the ecliptic'). The direction of this axis of rotation is not fixed in space; the attraction of the Sun and the Moon on the equatorial bulge of the Earth causes its axis of rotation slowly to precess—like that of a spinning top—in a period close to 25 800 years. This motion (responsible for the 'precession of the equinoxes' of our equatorial coordinate system) is the reason why the 'tropical year' (i.e. a time-interval between two successive passages of the Sun through the 'vernal point' of the intersection of the ecliptic with the equator) takes 365·242 20 mean solar days, or longer by 20 minutes and 23 seconds than one sidereal year.

For a long time the period of rotation (i.e. the length of the sidereal day) had been regarded as the master clock measuring the march of cosmic time. It is no longer used for such a purpose, for astronomical observations of the past 180 years have detected the existence of irregular variations in the length of the sidereal day amounting to as much as five parts in 10^8. This is probably a result of occasional small mass displacements in the interior of the Earth which are needed to keep its rotational momentum constant. Since 1967, the burden of keeping correct time has been lifted completely from the motion of the Earth or any other celestial body, and transferred to certain atomic vibrations of caesium-133 isotope which can be measured by laboratory techniques with a precision far exceeding that of any astronomical measurements.

The present length of the day on the Earth is close to 24 hours, but has this been so during the entire long past of our planet? In recent years interesting palaeontological evidence has been brought to light of the diurnal as well as the monthly effects on the growth of corals and certain fossil crustaceans of the Palaeozoic age. The available evidence appears to indicate that in the Cambrian period (some 500 million years ago) the days were substantially shorter, about 21 hours, indicating more rapid rotation of the planet. There were then about 415 days a year, and a synodic month lasted $31\frac{1}{2}$ days. In the Devonian period (some 380 million years ago) the length of the day increased to 22 hours and a month diminished to $30\frac{1}{2}$ days; and in the Upper Carboniferous (290 million years ago) the day increased to 22·6 hours, there being then 30·1 days in a month. By the end of the Mesozoic era (Upper Cretaceous period, some 70 million years ago) the day increased further to 23·67 hours (370·3 days in a year, and 29·9 in a month), until the present values of a 23·93-hour day and a 365·26-day year of 29·53-day months were established in the more recent past.

The reasons for this gradual change in the length of the day and the month are connected with a tendency for the Moon to slow down the axial rotation of our planet through tidal friction. The momentum thus lost to the Earth is transferred to that of the lunar orbit around the Earth, which is secularly increasing in size. The final stage of this evolution will be attained when our day on the Earth becomes equal to the length of the month—both about 60 days of our present time. The distance of the Moon from us will have increased to about 1·6 times its present value. At that time the recession of the Moon from us will ultimately cease, and no lunar tides will disturb our oceans any more. Several thousand million years have yet to elapse before this final stage of the evolution of the Earth–Moon system is attained.

Mass

Just as the Earth's size was used to serve as a basis for a triangulation of cosmic distances of increasing length, the terrestrial mass—once determined—provided a clue to the masses of all other celestial bodies. Needless to say, no basis existed for such measurements until it became possible to relate the mass of a body to a force exerted by its attraction, and this had to await the advent of Newton's theory of gravitation towards the end of the seventeenth century. In more specific terms, the discovery of the universal attraction, diminishing with the square of the distance of attracting bodies, enabled Newton to relate the magnitude of the terrestrial mass $m_{\oplus}$ to the acceleration g of a particle falling towards it by a simple equation of the form

$$g = \frac{Gm_{\oplus}}{r^2_{\oplus}},$$

where $r_{\oplus}$ denotes the distance of the falling body from the Earth's centre, and G the constant of gravitation.

The value of g could already be measured with fair accuracy in Newton's time, and the dimensions of the Earth were likewise known (from 1671) with adequate precision. However, the value of the constant G was not known to Newton, nor indeed for some time afterwards. Newton himself only guessed at it by assuming that the mean density of the Earth was between five and six times that of water*—a remarkably lucky estimate considering the fact that nothing was known in Newton's time about the state of the Earth's interior, and that common rocks found on the Earth's surface possess densities between 2800 and 3300 kg m^{-3}. Once we assume a given density for the Earth, the known dimensions of its globe permit us to evaluate $m_{\oplus}$ and, consequently, G from known values of g and $r_{\oplus}$. This Newton did; and from a value so obtained he went on to estimate the absolute masses of other planets as well as that of the Sun.

The first terrestrial determination of G was carried out in 1772 by Maskelyne in the Scottish Highlands. In the nineteenth century G was deter-

* Verisimile est quod copia materiae totius in Terra quasi quintuplo vel sextuplo major sit quam si tota ex aqua constaret (*Principia*, liber III, propositio 10).

mined much more precisely by laboratory experiments with torsion balances. Today it is known to be equal to $6 \cdot 67 \times 10^{-11}$ m^3 kg^{-1} s^{-2}. If we combine it with the observed gravitational acceleration $g = 9 \cdot 82$ m s^{-2} of free fall at a mean distance $r_\oplus = 6371$ km from the Earth's centre, the mass $m_\oplus$ of the Earth necessary to produce the requisite acceleration turns out to be $5 \cdot 98 \times 10^{24}$ kg. This corresponds to a mean density ρ_m of the terrestrial globe of 5520 kg m^{-3}—well within the limits estimated by Newton 300 years ago.

Density and composition

How is this mass distributed within the Earth? The fact that the Earth's mean density proves to be almost twice that of the rocks we find on the surface makes it evident that our Earth is not homogeneous. Its denser material must be confined to a zone close to the centre. The extent to which this is true was first indicated by the amount of flattening of the Earth at the poles by centrifugal force. This phenomenon was discovered from the changes in the rate of swing of a pendulum at different latitudes (caused by a change in local gravitational acceleration) by Richer and his colleagues in 1671–2 on an expedition to triangulate the distance to Mars between Cayenne (in French Guiana) and Paris. This is a well-known episode in the history of astronomy, and so is Newton's correct interpretation of the phenomenon as being due to the spheroidal shape of our planet. Now, the extent to which a rotating globe yields to centrifugal force to become flattened at the poles depends in a known way on its internal structure. The ratio of the centrifugal force to the observed flattening, which at the poles amounts to 22 km, indicates that the Earth's central density ρ_c is equal to three times its mean density ρ_m i.e. that ρ_c is approximately 17 000 kg m^{-3}. In other words the density of the material in the central parts of the Earth greatly exceeds that of iron (at normal pressure) and approaches that of gold.

More detailed information about the Earth's interior can be obtained from a study of seismic records of earthquakes, of which many thousands have been recorded in different parts of the world. Virtually all earthquakes observed so far originate in slips or other tectonic disturbances of the Earth's crust at relatively shallow depths below the surface; for more than 90 per cent of events the epicentres are less than 100 km deep. However, earthquake waves can often be recorded at the antipodes, and seismic records from distant places can disclose the extent to which the waves have been modified by passage through the intervening layers.

What do seismic studies reveal about the nature of the Earth's interior?

The main result has been a realization that the Earth as a whole consists of two distinct parts: a *core* whose radius of 3470 km is a little more than half (54 per cent) of that of the Earth is surrounded by a *mantle* making up the rest of the planet (see Figure 3). The volume of the core therefore represents

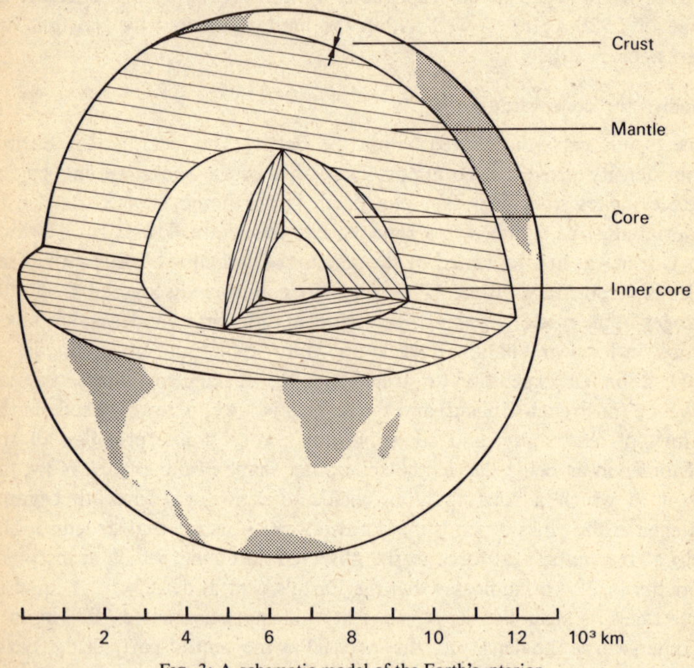

FIG. 3: A schematic model of the Earth's ınterior.

only about 16 per cent of that of the Earth as a whole, 84 per cent being occupied by the mantle, but in its structure and composition the core is so distinct from the mantle as to constitute almost a 'planet inside a planet'.

A more detailed analysis of the elastic properties of the Earth, as deduced from the available seismic evidence, has disclosed a number of other characteristics of the interior. Thus the crust of the Earth, supporting the land masses of the continents and the ocean floors, extends downwards to only granitic rocks. At a mean depth 33 km below the surface—where the weight of the overlying strata gives rise to a pressure of 9×10^8 N m^{-2} or 9000 atmospheres—the density increases rather suddenly from 2800 kg m^{-3} to more than 3300 kg m^{-3} (corresponding to that of basaltic rocks). It increases thereafter by self-compression throughout the mantle up to about 5700 kg m^{-3} at its base, where the pressure has risen to over a million atmospheres. So far as the composition is concerned, indications are that the outer part of the mantle (down to a subsurface depth of about 400 km) consists of rocks whose principal chemical constituents are oxygen, silicon, and aluminium; at greater depths aluminium may be gradually replaced by magnesium.

At the interface between the mantle and the core—some 2900 km below the Earth's surface—the density appears to jump suddenly from 5700 to 9400 kg m^{-3} at a pressure of $1{\cdot}4 \times 10^{11}$ N m^{-2}. So large a discontinuity strongly suggests that the solid mantle and fluid core differ not only in the physical state of their material but also in their chemical composition. In point of fact, a density of 9400 kg m^{-3} or more can be naturally accounted for only if we assume that the composition of the core is essentially *metallic*, its principal constituents being iron and nickel, with some admixture of molten silicate rocks.* Although the core occupies only 16 per cent of the Earth's volume, because of its relatively high density it represents over 31 per cent of the Earth's mass. Moreover, there are indications that inside this core (which behaves as a fluid) there is an inner core with a radius of some 1250 km. At the centre of this inner core the density of the constituent material has reached 17 000 kg m^{-3} and the pressure exceeds 3·6 million atmospheres. The inner core may consist of pure metals that are in the solid state because of the enormous pressure.

Iron, stored mainly in the core, accounts for almost 39 per cent of the mass of the Earth; oxygen, the second most abundant element by mass, contributes about 27 per cent. Silicon accounts for 14 per cent; magnesium 11 per cent; sulphur and nickel about 2·7 per cent each; and aluminium, with calcium, about 1 per cent. All the other elements together add less than 2 per cent to the total mass, of which 31·4 per cent is in the core, 68·1 per cent in the mantle, and 0·5 per cent in the crust.

* Some scientists have conjectured that the material of the core represents a high-density modification (metallic phase) of compressed silicates similar to those constituting the mantle. According to this view, the material of the core and the mantle is essentially the same, but its density jumps up discontinuously when a certain threshold pressure is attained at the interface between the two regimes. There is so far, however, no independent proof that this actually occurs in nature—inside the Earth or in the laboratory.

Internal temperature

Temperatures inside the Earth are difficult to determine by observation. Borings in the crust down to a depth of 2–3 km have indicated that the temperature in the mantle increases inwards at a relatively rapid rate of about 30°C per kilometre—attaining the boiling-point of water at an average depth of little more than 2·5 km. If surface water seeps down through occasional cracks to such a depth, it is brought to boiling point and ejected as hot steam under its own pressure as the hot springs or geysers that are familiar in many parts of the world.

If this inward rise in temperature continued at the same rate beyond the explored regions through at least the outer part of the terrestrial mantle, temperatures at which rocks melt (about 1500°C) would be reached at a subsurface depth of only 50 km. This is comparable with the depth of the combustion chambers of most terrestrial volcanoes. Volcanic eruptions may indeed represent nothing but occasional outbursts of deep-seated 'geysers' of lava and stones ejected from subterranean 'pressure-cookers'.

The actual source of the heat that accumulates in volcanic chambers may be partly mechanical (from frictional dissipation of motion into heat, for example), but the origin of the general source of heat that keeps the interior of the Earth at a relatively elevated temperature is chemical rather than mechanical. At the time of its formation, the Earth may have acquired some primordial heat of condensation but this could not account for its present temperature. The main sources of internal heat are the traces of certain radioactive elements, which produce heat by their spontaneous disintegration at a slow but steady rate. The principal radioactive elements occurring widely in nature that are known to emit measurable amounts of heat are uranium-235 and -238, thorium-228 and, among the lighter elements, potassium-40. All these play an important role in geophysics.

The amounts of these elements in common rocks of the terrestrial surface are, to be sure, minute. Granites and basalts contain from one to ten parts per million (ppm) by weight of uranium, which has a half-life* of $4{\cdot}5 \times 10^9$ years and decays spontaneously into lead. Thorium, with a half-life of about $1{\cdot}4 \times 10^{10}$ years, constitutes about 0·04 ppm of the Earth's crust and also decays into lead. In so doing these elements keep releasing minute amounts of heat equal to about $1\text{–}2 \times 10^{-10}$ kJ kg^{-1} (see Table 6).

This heat generation represents a tiny trickle. It does, however, act dependably and continuously and the temperature to which it can eventually give rise depends on the balance between heat generation and loss. Heat is produced throughout the entire volume of the Earth (provided that the radioactive material is uniformly distributed), and is lost by conduction (or radiation) through the surface. An analysis of the heat balance discloses that if the mass of the Earth contained as much uranium or thorium throughout its interior as we find on the surface, the amount of radiogenic heat produced in the course of a few billion years and increasingly bottled up in

TABLE 6
Radioactive disintegrations used in planetary chronology

Disintegration chain	*Emission**	*Half-life (in 10^9 yrs)*	*Total heat production (10^{10} kJ kg^{-1} of the mother substance)*
Uranium-238–lead-206	$8\alpha + 6\beta$	4·51	1·922
Uranium-235–lead-207	$7\alpha + 4\beta$	0·713	1·856
Thorium-232–lead-208	$6\alpha + 4\beta$	13·9	1·655
Rubidium-87–strontium-87	β	49·8	0·00
Potassium-40–argon-40	e	1·306	0·171

* $m\alpha$ indicates a total loss of m α-particles in the respective chain; $n\beta$, that of n β-particles; and e, the capture of an electron.

the interior of our planet (because its escape by conduction is a very slow process) would raise the internal temperature far beyond the melting-point of its constituent material not only in the core, but also throughout most of the mantle.

The seismic observations of the shear waves propagating freely through the mantle demonstrate that this cannot be the case, for shear waves cannot be transmitted through a liquid. We are therefore driven to the conclusion that the crust has been enriched with radioactive material, far beyond the level prevalent in the deep interior, by nuclear processes that accompanied the last formative stage of our planet. As we shall see later, the same appears to be true of the Moon. When one takes into account the effect of pressure on the melting-points of terrestrial materials, it becomes apparent that the temperature prevailing at the interface between the mantle and the core cannot exceed about 4000°C. Near the Earth's centre the temperature may attain 5500–6000°C—values not very different from the effective temperature of the solar surface—but the pressure there is so high (and increases more rapidly than the temperature) that the metallic material of the inner core may again behave as a solid. It is only the material of the core proper, comprising a shell with radii of 1250 and 3400 km, that behaves like a liquid. Thermal convection currents inside the conducting material of this shell give rise to a self-excited dynamo that is responsible for the maintenance of the Earth's magnetic field.

Whatever temperatures now prevail in the Earth's interior, they cannot remain constant but must change in the course of time. A warm Earth must be constantly redistributing its internal heat supply by conduction and must lose some heat by radiation into space, while new heat is being produced by spontaneous disintegration of radioactive elements in the interior at a known rate. If this radiogenic heat comes chiefly from long-lived elements like uranium, thorium, or potassium (with half-lives of the order of 10^9 years), it follows that the Earth must at present still be warming up (albeit at a very slow rate) and was internally cooler in the past. If so, however, it is possible that its metallic core did not exist in the primordial Earth, but that its iron and nickel were extracted only gradually from the rocks, in which they occur in small amounts, by the rising temperature. The newly-formed Earth need not, therefore, have possessed the same degree of differentiation in internal structure as it does today, nor need its present profile represent the final stage of its evolution.

The Earth's crust

The crust of the Earth, extending not more than 30–40 km down below the surface, constitutes a shell whose mass is less than 10^{-4} that of the Earth as a whole. It consists of rocks whose average density ranges from approximately 2800 kg m^{-3} in the continental land masses to 3300 kg m^{-3} in the ocean floors and underlying rocks. The principal chemical constituents of these rocks are oxygen (about 47 per cent by weight), silicon (28 per cent), and aluminium (8·4 per cent), followed by iron (2·5 per cent), calcium (2·4 per cent), and other elements in diminishing amounts.

The continental areas are, broadly speaking, mainly of granitic composition (quartz, orthoclase, mica, etc.) and the underlying layers basaltic (olivine, plagioclase, pyroxene). The former are less dense, and float on the underlying basaltic substrate in accordance with the principle of hydrostatic equilibrium (isostasy). They can also move both vertically and horizontally on the underlying substrate if impelled to do so by forces operating in the interior. A post-glacial uplift of Fenno-Scandia (at a rate of almost half a metre a century) is the result of the unloading of extensive glaciers that covered the area during the last ice age. Moreover, different lines of geophysical evidence disclose the probability of slow lateral motions in the

Earth's crust. This continental drift is believed to operate on a time-scale that could have profoundly altered the face of the Earth several times in the course of its existence. It seems, indeed, that not more than 10^9 years ago the Earth could have possessed only one continuous continental area, surrounded by a global ocean on all sides, and its diversification into the present five continents and smaller islands may have occurred since that time.

Before we consider the nature of the forces that could have split up the primordial crust and sent its fragments floating in different directions on the surface of our globe, let us inquire about the *age* of the rocks constituting them: i.e. the time that has elapsed since their solidification. As has been known from the beginning of this century, the ages of rocks can be ascertained from the measured progress of spontaneous disintegration of certain radioactive elements. A list of the principal nuclear clocks used for this purpose is given in Table 6, which lists the mother and daughter products and their half-lives. The half-lives range from $0{\cdot}713 \times 10^9$ years for the uranium-235 → lead-207 distintegration to $47{\cdot}1 \times 10^9$ years for the rubidium-87 → strontium-87 β-decay. But they all possess one feature in common: they start marking the time from the moment when the rock containing them has solidified. It is as if the dials of the radiometric clocks (which measure the proportion of the daughter product to the mother substance) are automatically set back to zero whenever the respective mineral has been remelted—just as (on a time-scale six orders of magnitude shorter) the well-known carbon-14 method of dating can tell the archeologist the age of any organic tissue, measured from the moment when the organism stopped its intake of carbon-14 from the atmosphere by breathing at the time of its death.

When the uranium-lead, rubidium-strontium, and potassium-argon dating methods were invoked to determine ages of terrestrial rocks, it was found that the oldest rocks found anywhere on the surface of the Earth are no more than $3{\cdot}5$–$3{\cdot}6 \times 10^9$ years of age, and at least 10^9 years younger than the oldest known meteorites (cf. Chapter 10) or lunar rocks (Chapter 7). If we assume—reasonably enough—that the Earth as a whole is about as old as the oldest solid particles found anywhere in the solar system (i.e. $4{\cdot}5$–$4{\cdot}6 \times 10^9$ years), the gap in age discloses that the formation of the Earth's planetary crust must have been a very slow process.

The only reasonable explanation of the lower ages of the terrestrial rocks would seem to be that the rocks now forming the crust of our planet were not a part of its original surface, but emerged from the interior at a later date. We are led to such a conclusion by the fact that large parts of the terrestrial surface—and not only of its continental areas, but also of its ocean floors—appear to be relatively young, with radiometric ages of not more than a few hundred million years.

The physical mechanism that could be responsible for these facts may be slow thermal convection currents, driven by the heat engine in the terrestrial mantle. Seismological evidence has amply demonstrated that for disturbances of short duration the material of the Earth's mantle behaves like an elastic solid. However, the observed extent of the polar flattening of our planet also demonstrates that the Earth responds to a continuous stress like a fluid. Since the time-scale of the postulated mantle convection is very long, slow convection remains at least a theoretical possibility. Its existence would enable us to kill two birds by one stone: to provide a motive power for continental drift, as well as for a thermal consumption of surface rocks which may at times be drawn deep enough into the mantle to be melted and their nuclear clocks reset.

The time-interval at which this seems to have recurred appears to be generally of the order of a few hundred million years. Only in rare parts of the terrestrial surface—where there has been no convection motion—can rocks be found that are substantially older. Such regions appear to be the Canadian shield and the central and South African plateau, where rocks 3500 million years old have been discovered. Such rocks cover no more than a few per cent of the terrestrial surface; the rest is cosmically much younger.

Another by-product of such a process could also be the formation of continental mountain chains such as the Alps, Himalayas, or Andes, some of which extend for thousands of kilometres. Such mountains (not only those parts of them that loom above the surface but also their subterranean roots) are known to consist of relatively light, mainly granitic, rocks that are characteristic of the continental crust. If a continental land mass—thousands of kilometres in extent, but not much more than 20–30 km deep—floats on the denser substrate like a leaf carried by a slow stream the thin continental layer may not always move as a rigid block, but may warp up along lines where the crust fails to yield to lateral pressure.

This may indeed be the origin of folded mountain chains, so typical of the terrestrial landscape (and, as we shall see below, conspicuous by their absence on the Moon). Typical examples are like the great chain commencing in the Bay of Biscay with the Pyrenées, Alps, and Carpathians in Europe and continuing through the Caucasus to Pamir and the Himalayas in Asia; or the great chain of the Andes and Rocky Mountains running from Patagonia to Alaska along the entire west coast of the American continent.

Geologists studying the detailed anatomy of such formations tell us that the mountains just mentioned (with a host of less conspicuous links) originated by folding of the crust in the relatively recent past (some 30–50 million years ago) as a part of a great orogenic process that occurred during the Tertiary era. Similar processes have operated before in the Earth's history—for example, in the Permian and Devonian periods of the Palaeozoic era some 250 and 400 million years before our time. Palaeozoic mountain chains (such as the Caledonian mountains in Scotland, or the Appalachians of the east American coast) loom very much less conspicuously on the maps because erosion has largely obliterated them in the course of their long past.

Oceans

More than 70 per cent of the surface of the Earth is covered by water and the oceans are a characteristic (and unique) feature of the face of our planet. The ocean basins represent only shallow pockmarks, but their maximum depth below sea-level (close to 11 km) is more than 2 km greater than the altitude of the highest mountains on land (8884 m). Ten kilometres represent, we must remember, less than one part in 600 of the terrestrial radius, and less than one-half of the polar flattening produced by the daily rotation of the Earth.

When examined in detail, the ocean floors are far from smooth. If all water were to disappear from the Atlantic, we should find on its floor the largest mountain chain on the Earth, the famous mid-Atlantic ridge. This runs in a north–south direction continuously for some 10 000 km. The highest mountains on the Earth are not in fact in the Himalayas, but on the floor of the Pacific. The Hawaiian volcanoes Mauna Kea and Mauna Loa attain altitudes of only 4208 and 4170 m above sea-level, but they rise from an ocean floor 5800 m deep. Their actual altitudes above the surrounding abyssal plain are therefore close to 10 000 m, more than 1100 m higher than Mount Everest.

The total mass of the water constituting the terrestrial *hydrosphere* is about $1{\cdot}4 \times 10^{21}$ kg, or a little more than one ten-thousandth of the mass of the Earth as a whole. If this water were distributed uniformly all over the globe, it would cover the Earth with an ocean about 1800 m deep. Chemically, ocean water is far from pure: each cubic kilometre (weighing 10^9 tons) contains, on the average, 19 million tons of dissolved chlorine, 10·6 million tons of sodium, 1·3 million tons of magnesium, and proportionally smaller amounts of other elements* which make it truly salt water—chemically as well as by taste.

* Among which we should find about 300 kg of silver and 4 kg of gold.

Where did all this water come from? We are fairly sure that the newly-formed Earth contained no water on its surface. It was born dry and its oceans have since exuded from the interior by thermal 'cracking' of the hydrates. The water molecule is a cosmically common and very stable product of nature. We find it (by its radio-emission) to exist in interstellar space in the form of tiny ice crystals, and we know of many minerals on the Earth, such as obsidians and many other kinds of volcanic glasses, which contain as much as ten per cent (by weight) of water molecules imprisoned in their structure. At the normal temperatures that prevail on the surface of the Earth such hydrates are, in general, stable, and can retain their water for an indefinite length of time. Moderate heating (to temperatures between 500° and 1000°C) is, however, sufficient to break the chemical bonds that keep the water molecules within the molecular structure of a hydrate, and to expel it in the form of a steam which can subsequently condense. We have mentioned above that temperatures of this order are likely to have been exceeded by now in most parts of the Earth's interior. If so, however, the hydrates there are likely to have lost most of their water. If water was originally present in the primordial material from which the Earth was formed to an extent of 1 part in 10 000 or more, thermal cracking of the hydrates and the outward seepage of the free water as superheated steam could eventually have produced all the water we find in all the oceans of the Earth.

How old are the oceans? While nothing is known about a possible hydrosphere of our planet in the first 'dark aeon' of its existence, the oldest preserved strata (more than three thousand million years old) bear evidence of the existence of 'sediments' which required water for the formation of their deposits. Moreover, the original juvenile water squeezed out by heat from the Earth's interior probably contained few salts dissolved in it before it emerged to the surface. Chemists estimate that, in order to acquire its present salinity, sea water must have been on the surface in the liquid state for not less than three thousand million years.

The atmosphere

The continents and the hydrosphere thus form the topmost layer of the Earth's crust, constituting a skin whose depth extends barely to two parts in a thousand of the terrestrial radius. Surrounding this surface is the gaseous envelope known as the *atmosphere.* That a terrestrial planet of the mass of the Earth should possess an atmosphere is not accidental, but is a consequence of the same processes that have endowed the Earth with its hydrosphere. If the latter originated by a 'de-fluidization' of the Earth caused by a gradual build-up of radiogenic heat in its interior, the atmosphere originated by its degassing—through the liberation from the interior of volatile elements whose molecular weights were large enough to prevent their escape from the Earth's gravitational field. The gradual formation of the hydrosphere and atmosphere represents therefore two different aspects of the same process. Moreover, an atmosphere capable of exerting adequate air pressure is necessary for the maintenance of any liquid on the surface. Planets can exist that possess atmospheres but no hydrospheres (Mars and Venus are examples), but the converse is physically impossible.

The chemical composition of our atmosphere is well known, and is in agreement with our expectations of the ability of the Earth's gravitational field to retain gases of given molecular weight. The principal constituents are nitrogen (75·5 per cent by weight) and oxygen (23·1 per cent), followed by argon (1·3 per cent), neon (13 ppm), helium (0·7 ppm), and diminishing amounts of heavier inert gases (krypton, xenon), present in quantities so minute that a chemist would describe them as impurities. Variable amounts of water vapour (0·01–0·1 per cent) and carbon dioxide (0·03 per cent) make up the rest. Hydrogen, a principal constituent of water in the hydrosphere, is too light to be permanently retained as a gas in the atmosphere, and it is present (in the outer layers) in barely significant amounts. Apart from water vapour and carbon dioxide, the composition of the atmosphere is remarkably uniform to a height of at least 100 km. This testifies to the efficiency with which atmospheric gases with different molecular weights are intermixed by atmospheric turbulence.

The total mass of the terrestrial atmosphere is $5{\cdot}3 \times 10^{18}$ kg. It constitutes, therefore, only about one-millionth of the mass of the Earth as a whole, and less than 0·3 per cent of that of its hydrosphere. An atmospheric column of 1 m^2 cross-section at sea-level weighs about 10 340 kg, and exerts an average pressure of 10^5 N m^{-2} (one bar), which is capable of balancing a column of mercury about 760 mm in height. The volume of space occupied by the atmosphere is, however, more impressive, extending as it does to an altitude of several hundred kilometres. The bulk of the atmosphere is, to be sure, confined (by self-compression) to its base, which the meteorologists describe as the *troposphere*. This extends to a mean altitude of eleven to fourteen km (according to geographic latitude). The highest mountains on the Earth just approach the top of this layer, and modern jet aircraft skirt it in cruising flight. Above lies the *stratosphere*, so called because the decrease in barometric temperature throughout the troposphere becomes arrested (at some 218 K); at higher, stratospheric, levels the temperature actually begins to increase.

At altitudes of 60–80 km we encounter another distinct layer, the *ionosphere*. At these heights the ambient air pressure diminishes to only about one part in 5×10^5 of that at sea-level, and the air density (which at sea-level is close to $1{\cdot}2$ kg m^{-3}), is only about 9×10^{-5} kg m^{-3}. Air so rarefied is easily ionized by energetic radiation from the Sun, and the number of electrons at this height may amount to as many as 10^{12} m^{-3}. Such a layer, extending almost up to 300 km (and itself made up of several layers) acts like a metallic mirror or mesh that can reflect high-frequency radio waves downwards—and can back-scatter similar waves reaching our Earth from space.

The height of the ionosphere, as well as its transparency, oscillates between day and night, and its outer surface may attain altitudes of nearly 300 km. The temperatures prevalent at these levels vary, and may attain values far in excess of those encountered on the surface. The density of such a hot gas ranges, however, only between 10^{-8} and 10^{-11} kg m^{-3}. These regions can give rise to the beautiful displays of polar aurorae; and even in their lower layers the density is sufficient to burn up the majority of meteors by atmospheric resistance (see Chapter 10).

Above the ionosphere, the remaining air becomes so rarefied, and the mean free path of the individual particles so long, that we are dealing, not with a gas, but with an assembly of individual particles that describe ballistic trajectories of increasing length between individual collisions which become progessively less frequent. This is the terrestrial *exosphere*, where our planet borders on interplanetary space. Some of the more energetic particles are lost to space when their speeds exceed the escape velocity from the Earth's gravitational field.

In another sense the astronomical domain of the Earth does not end at the neutral exosphere; for far above it we encounter condensations of charged particles (mainly protons and electrons) trapped by the magnetic field of the Earth, and known (after their discoverer) as the terrestrial van Allen belts, or, more generally, the terrestrial *magnetosphere*. These form a doughnut-shaped ring around the Earth (Figure 4), which is populated by charged particles of external (solar, and partly interstellar) origin.

Theoretical studies of the motion of charged particles in the vicinity of the terrestrial magnetic dipole were initiated many years ago by Störmer, who pointed out the discrete nature of the different belts that the charged particles of external origin may, or may not, enter. Possibly the first suggestion that the inaccessible Störmer regions may actually be populated came from Alfvén in 1947. The existence of actual corpuscular radiation trapped in the inner region had already been indicated by experiments with the Russian

Sputnik 2 in November 1957. Confirmation had, however, to await the U.S. Explorer 1 as the first man-made probe that actually penetrated the interior of the first (close) radiation belt in the early part of 1958, and the correct interpretation of its data was advanced by van Allen in May of that year.

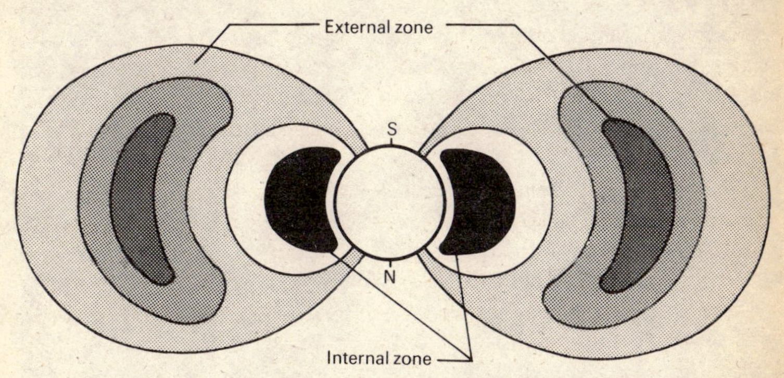

FIG. 4: Geometry of the van Allen belts of geomagnetically trapped charged particles around the Earth.

A schematic anatomy of the van Allen belts is illustrated in Figure 4. The inner ring (shown in cross-section as two lobes) extends radially from about 1000 to 10 000 km above the Earth's surface, and contains a relatively large proportion of high-energy protons. The outer ring is much more voluminous (extending from some 25 to 60 thousand km in the radial direction) and less densely populated. It contains mainly electrons moving about more slowly, but still possessing energies far exceeding the thermal energies of particles in the exosphere. Both the protons and electrons orbiting (or librating) in the terrestrial dipole field come largely from the solar wind, though protons and electrons in the inner belt may also be produced by interactions between molecules in the extension of the terrestrial exosphere and cosmic rays.

The particle flux inside both belts fluctuates with solar activity (being of the order of 10^7 particles per m^2 per steradian* in the outer belt, and exceeding 10^{10} particles in the inner). The geometrical cross-sections of both belts, and especially of the exterior one, vary considerably in the course of time. Their regions are in fact not sharply defined at all, for trapped corpuscular radiation must be spread throughout the Earth's magnetic field. The inner and outer van Allen belts should be thought of merely as regions where the space density of moving charged particles becomes greater than elsewhere.

PLATE 5: The Earth from Space—photographed from an ATS–III satellite at a distance of 35 700 km. The South American Continent and Atlantic Ocean are seen in front through a partial veil of clouds, while West Africa protrudes near the limb*.

The existence of the van Allen belts cannot be detected from the surface of the Earth by any kind of optical observation, for the amount of sunlight scattered by their particles is too small to give rise to a visible glow. The charged particles constituting them pose, however, a definite hazard to any kind of spacecraft traversing them. Fast-moving particles striking its walls produce X-rays, which are dangerous to man and may interfere with scientific experiments aboard. In order to escape injury, space travellers must either pass through these belts very rapidly (so as to be exposed to the harmful radiation for as short a time as possible), or leave the Earth by a polar route, for van Allen belts do not extend over the polar regions (see Figure 4).

This completes a brief survey of the anatomy of our planet. We have found the Earth to consist of a (probably) solid inner core of nickel-iron, with a radius of 1250 km, surrounded to 3470 km by a liquid shell consisting of much the same material (with possibly some admixture of molten silicates), and overlain further by a mantle of silicate composition, covered by a crust. This crust supports the continents and the oceans (the hydrosphere), and is protected by a gaseous atmosphere of several distinct layers, gradually petering out through an exosphere into a magnetosphere, which represents the physical boundary of our planet. Although the solid globe has a mean radius of only 6371 km, the limits of our magnetosphere may extend at times almost ten times as far from the Earth's centre in the direction of its magnetic equator, and it is not until we reach this point that we have completed the physical portrait of our mother-planet (Plate 5).

4

The Planet Venus

IN THE PRECEDING chapter we discussed the principal astronomical and physical properties of the prototype of the terrestrial planets—our Earth. We must now pay courtesy calls on the other members of this little planetary family and, in particular, on our closest planetary neighbours Venus and Mars.

The most important kinematic and physical characteristics of different members of this planetary group, ranging in mass from the Earth to its Moon, have already been listed in Tables 4 and 5. A glance at these discloses at once that the planet most like our Earth in its vital statistics is Venus.

This second-innermost planet of the solar system revolves around the Sun in 224·7 days in a mildly eccentric orbit (of eccentricity $e = 0{\cdot}006\ 79$) inclined to the ecliptic by only $3°23'40''$, which brings it at inferior conjunction within $40{\cdot}7 \times 10^6$ km of the Earth. At such times, it becomes our nearest neighbour, though still more than a hundred times as far away as our Moon. The planet Mars never approaches us within less than $55{\cdot}5 \times 10^6$ km. A spacecraft, which can nowadays reach the Moon after a journey of 60–70 hours, must spend not less than 3 months on its way to Venus; and even light or electromagnetic signals sent out from the Earth will take 140 seconds to reach it. On the other hand, at superior conjunction (when the planet is on the other side of the Sun) the distance increases to 258×10^6 km or 14·3 minutes of light travel.

Between these distances, the apparent diameter of Venus (as seen from the Earth) ranges from $10''$ at superior conjunction to $64''$—more than one minute of arc—at the time of closest approach; this corresponds to a globe of radius little less than 6100 km, not deviating significantly from a sphere. The surface of Venus is, therefore, equal in size to 90·6 per cent of that of the Earth, and its volume to 86·2 per cent of the terrestrial volume.

The mass of this planet, like that of any other celestial body, can be determined only from observable effects which its attraction can exert on another body to perturb its motion. The traditional methods have involved perturbations caused by Venus in the motions of the Earth or Mercury. Recently, however, this role has been wholly taken over by the deep-space planetary probes which have been sent out since 1961 to reconnoitre the environment of Venus at closer range. As a result of very accurate tracking of the motion of such spacecraft in the proximity of Venus, we know now that the reciprocal mass of the planet is given by

$$m_{♀}^{-1} = 408\,520 \pm 10,$$

i.e. a mass 408 520 times smaller than that of the Sun, and 81·50 per cent of that of the Earth, or $4{\cdot}87 \times 10^{24}$ kg. The mean density of the Cytherean* globe is, therefore, 0·945 times that of the Earth, or 5230 kg m^{-3}, and its gravitational acceleration on the surface is 0·9 of the terrestrial one, or 8·82 m s^{-2}. In these, and other, respects Venus would, therefore, appear to be almost a twin to our Earth. Yet appearances are often deceptive, as they have proved in the present case. For all the apparent superficial similarity there is a vast difference between other physical properties of the two planets—a difference that we did not begin to comprehend until the advent of radar and spacecraft astronomy.

Before that time, our knowledge of the more intimate properties of our sister planet was very limited. In 1611, Galileo Galilei recognized—and expressed in one of his concealing anagrams (*Mater Amorum aemulatur Cynthiae formas*)—that the telescopic image of Venus simulates the phases of the Moon. This fact had important cosmological consequences at the time. For, according to the Ptolemaic astronomy (in which both the Sun and Venus were supposed to revolve around the Earth) Venus could never show us more than one half of its illuminated surface (since, according to Ptolemy, the planet was always between us and the supposed orbit of the Sun). Therefore, when Galileo observed with its telescope that Venus exhibited a gibbous* phase as well as the crescent—even though it never deviates from the Sun in the sky by more than 48° at maximum elongation —this fact proved that Venus must revolve around the Sun rather than around the Earth. The establishment of this fact sounded the death-knell to the Ptolemaic system, although it did not completely vindicate the Copernican one; for it was consistent with either the heliocentric Copernican system, or the Tychonic system (according to which all planets except the Earth revolved around the Sun, but then together with the Sun around the Earth).

Whenever it is visible from the Earth, Venus appears brighter than any other object in the sky (other than the Sun or the Moon), ranging from −3·3 to −4·2 in stellar magnitudes. Since Venus revolves around the Sun in an orbit interior to that of the Earth, it is bound to exhibit phases in solar illumination much as the Moon does to us. As it moves from superior to inferior conjunction (see Plate 6), an increase in apparent diameter (due to increasing proximity) will at first more than make up for a diminution of the illuminated portion of its disc, and the planet grows brighter. As its crescent narrows further, however, the apparent brightness begins to diminish in spite of increasing proximity.

As a result of the combination of these effects, Venus reaches its maximum brightness not when it is nearest to us, but about 36 days before or after inferior conjunction*, at an elongation of 39° from the Sun (i.e. less than the maximum elongation of 47°·5), when it appears in the telescope like the Moon about two days before the 'first quarter'. At such times, Venus is bright enough to cast shadows at night, and can be seen with the naked eye in full daylight.

Such brightness of the planet discloses that its surface scatters incident sunlight effectively—approximately 59 per cent of it, in comparison with a mere 7 per cent scattered by our Moon—from which we conclude that the visible surface of Venus (the layer responsible for the scattering) must be much smoother than the surface of the Moon. Moreover, the colour of Venus is only slightly yellower than that of illuminating sunlight, so the scattering process must be almost independent of frequency. In other words, the Cytherean disc appears to be rather monotonously white—as might be expected of a continuous *cloud* layer completely covering the planet. That this is probably the case is also suggested by the fact that, unlike the Moon or Mars, Venus does not exhibit any permanent surface markings; the usual appearance of the planet shows a uniformly bright disc, shading off smoothly towards the terminator* without any safely recognizable detail.

The atmosphere of Venus

If the surface of Venus is perpetually covered by clouds, the planet must be surrounded by an extensive atmosphere; and telescopic observations indicating this have indeed been available for a long time. Near the time of inferior conjunction, the horns of the Cytherean crescent are often seen to extend noticeably beyond the sunrise terminator; and when the planet is very close to the Sun in the sky, the cusps of the crescent have been observed to coalesce in a complete aureola surrounding the whole disc. This phenomenon is due to diffuse reflection of sunlight in the planetary atmosphere (the same process

PLATE 6: The planet Venus at five different phases, photographed by H. Camichel with the 15-inch refractor of the Observatoire du Pic-du-Midi.

which produces our twilight) and from its intensity we can estimate the total amount of gas necessary to produce it: it is equivalent to a layer of about 1 km of gas at our atmospheric air pressure. This is, of course, the amount of gas above the top of the visible cloud layer; the method cannot tell what is below.

What does this atmosphere above the clouds consist of? As usual, we have to turn to the spectroscope for an answer. In 1932 the American astronomers Adams and Dunham of Mt. Wilson Observatory detected in the near-infrared part of the Cytherean spectrum three complex bands (with peaks at 782·0, 788·3, and 868·9 nm) due obviously to molecular absorption, which have no counterpart in the spectrum of illuminating sunlight; these were subsequently identified with the absorption spectrum of carbon dioxide. Moreover, from the intensity of these bands it has been inferred that the amount of carbon dioxide above the visible cloud surface must be equivalent to a layer not less than 400 metres thick at the Earth's atmospheric pressure, which is about 250 times the quantity of carbon dioxide present in the Earth's entire atmosphere.

Oxygen and water vapour could, in principle, be detected in the same manner for, unlike nitrogen or argon, the absorption of their molecules produces imprints in the visible part of the spectrum. However, all spectroscopic tests for oxygen have so far proved to be entirely negative, indicating a well-nigh complete absence of this gas in that part of the Cytherean atmosphere accessible to observation. It was not until quite recently (1966) that the first traces of water vapour were detected in the deep-red (wavelengths 818·9 and 819·3 nm) part of the spectrum of Venus by Spinrad and Shawl from Berkeley, and independently by Belton and Hunten at Kitt Peak, in amounts corresponding to not more than 10–20 microns of precipitable water. (Thus if all the water in the atmosphere were to precipitate as rain and cover the surface of the planet with a uniform layer, the layer would be 10–20 μm deep.) This is in agreement with previous measurements of the infrared H_2O absorption at 1130 nm carried out in 1964 by Strong from unmanned balloons at a height of 80 000 feet. The atmosphere of Venus above the visible surface appears, therefore, to be very dry—about as dry as our own terrestrial stratosphere above the high-flying cirrus clouds—and the reasons for this will emerge when we inquire into the temperature prevalent over the visible parts of the Cytherean surface.

The temperature of Venus

The temperature of any cosmic body can be deduced from the measured intensity of its thermal emission and, for solid planetary bodies, the bulk of this emission can be expected (in accordance with Planck's Law) to occur in the near infrared. When the first reliable measurements of infrared emission from Venus between wavelengths of 8–12 μm were performed in 1955 by Pettit and Nicholson, it transpired that while the mean temperature of the 'daylight' hemisphere was 235 K, that of the 'night' hemisphere was 240 K and therefore higher than the daytime temperature. The difference between the two is scarcely significant and all the measurements taken together indicate that the temperature prevalent above the visible surface of Venus remains virtually constant regardless of day or night.

How to explain this perplexing phenomenon? The correct clue was probably found by Strong, who pointed out that these measured temperatures are virtually identical with the Schaefer point, below which saturated water vapour cannot cool off without spontaneous crystallization. Therefore, the clouds on Venus are probably just like the high cirrus clouds in our own atmosphere—i.e. ice crystals—and the failure of their outer surface to cool at night is probably due to the liberation of the latent heat of water vapour (about $2{\cdot}4 \times 10^3$ kJ kg^{-1}) as the vapour condenses to form clouds. A recent identification of hexagonal ice-crystals in the Cytherean cloud layer by their optical 'halo' effect seems to have clinched this point beyond reasonable doubt.

How high are these clouds above the surface of the planet? This question obviously cannot be decided by observations made in visible or near-infrared light; for the cloud cover appears to be totally opaque to this light at all times. In order to penetrate deeper, light of much longer wavelength, which is not absorbed by water vapour, must be employed. This means that we must turn our attention to wavelengths in excess of one millimetre—the radio spectrum of the planet Venus. The exploration of this began only in the 1960s, and the results could not have been more dramatic. Whereas in the near infrared the intensity of the Cytherean thermal emission corresponds to a mean temperature of 238 K, the intensity at wavelengths of 8–9 mm proves to correspond to temperatures of 353 K; and, in the centimetre range, it corresponds to values close to 673 K or 400°C.

This latter temperature refers undoubtedly to the solid surface of the planet at the base of the atmosphere; for at centimetre wavelengths the atmosphere of Venus (like our own) becomes effectively transparent. If, however, the high temperature is due to the storage of sunlight by a 'greenhouse effect' which traps solar heat through absorption by its gaseous constituents, the amount of gas involved in the process must evidently be enormous. Calculations of Cytherean model-atmospheres have disclosed that, in order to account for a ground temperature of 400°C, an atmosphere consisting predominantly of carbon dioxide would have to attain a ground pressure in excess of 100 terrestrial atmospheres; and that the top of the cloud layer should be some 45–50 km above solid ground. Both these expectations have been borne out by subsequent spacecraft and radar work.

Thus it is not until altitudes which, on the Earth, would take us well above the ozone layer, that anyone emerging from the Cytherean atmosphere could get a glimpse of the starry sky. Far below the deceptive calm of these rarefied heights, clouds impenetrable to visible frequencies conceal a veritable inferno of shadowless semi-twilight, filled with hot carbon dioxide under enormous pressure. And yet even this inferno has already been penetrated by inanimate messengers, first in the form of radar signals, and later by actual spacecraft (Mariner, Venus).

The first terrestrial messenger to Venus was the Russian probe Venera 1, launched in February 1961. Although it failed to reach its goal it proved to be a harbinger of greater things to come, and paved the way for the U.S. fly-by Mariners 2 and 5 (which paid close calls on our sister planet in 1962 and 1965), the fly-by Venera 2 of 1966, and Veneras 3, 4, 6, and 7 which actually entered the Cytherean atmosphere and descended through it with the aid of suitable parachutes down to an increasing depth. Venera 7 of December 1970 actually touched the ground.

The scientific gains of these flights, telemetered to us on Earth across a space gap of many millions of kilometres, were far-reaching. The composition of the Cytherean atmosphere was found to consist of some 95 per cent of carbon dioxide, 2 per cent of nitrogen, and 1 per cent of water vapour, the bulk of the rest being (probably) neon and argon; but free oxygen seems conspicuous by its absence (its amount, if any, is less than 0·01 per cent). Moreover, magnetometers carried aboard several of these spacecraft failed to detect any planetary magnetic field of more than 0·01 gauss in strength; and the upper limit of a possible magnetic dipole of Venus was estimated to be no more than 0·002 of that of the Earth.

In a real sense, however, the space exploration of Venus commenced on 10 March 1961 when the first human contact with Venus was established by radar, before an approach of the first spacecraft. On that day, a radio signal beamed at Venus from the Jet Propulsion Laboratory of the California Institute of Technology at a wavelength of 12·5 cm produced a detectable echo within a few minutes of double-transit time; and, since that time further important contributions have been made by the Lincoln and Arecibo Observatories in the United States, as well as by the Radiophysics Institute of the U.S.S.R. Academy of Sciences in Russia.

The contributions of radar astronomy to the study of Venus since 1961 have been extensive and important, and some of them have been unique. By the determination of the time-lag between the outgoing and returning signal, the distance separating us from Venus has been obtained with a precision far surpassing that of any previous astronomical triangulation; and radar measurements at different times (i.e. when the planet occupies different positions in its orbit around the Sun) have led to a determination of the orbital elements of the planet so precise as to mark an entirely new epoch in celestial astronautics. For instance, the semi-major axis of the Cytherean orbit ($0{\cdot}723\,329\,860 \pm 0{\cdot}000\,000\,002$ AU) is now known to us within an error of two parts in 10^9, and the length of the astronomical unit (i.e. the semi-major axis of the terrestrial orbit around the Sun) was deduced by the same method to be 149 597 892 km in length (or 499·004 786 light seconds) with an error of only ± 5 km (or 5 microseconds of light). This value is more than 2000 times as precise as the best previous 'astronomical' determinations of this unit by celestial triangulation, and overwhelmingly demonstrates the great progress made by these new methods.

The strength of the radar echoes from Venus also furnishes valuable information on reflecting properties (i.e. the dielectric constant or permittivity) of the Cytherean solid surface. The average dielectric constant of Venus so obtained turns out to be about four to seven times as large as that of a vacuum—in contrast with a figure of two for the lunar surface, or about fifteen (average) for the Earth. The Cytherean value of the dielectric constant is definitely smaller than it is for the Earth (indicating a smaller proportion of solid rocks in the radar-reflecting medium). It is comparable with that of dry, sandy, desert ground but, on the other hand, it is indicative of a more compact surface than that of the Moon.

The rotation of Venus

The greatest contribution of radar astronomy to the study of Venus has, however, been the determination and measurement of the axial rotation of the planet. As we have already mentioned, an impenetrable layer of high clouds permanently prevents us from observing any markings on the surface whose motion could indicate the length of a Cytherean day. Spectroscopic measurements of radial velocities with different positions of the slit on the planet's disc suggested only that the axial rotation of Venus—if any—must be very slow. On the other hand, the radar echoes at the centimetre wavelengths—our only permanent contact with the planetary surface so far—should clearly have their profiles influenced by planetary spin. Not only do different parts of the surface return signals with a different time-lag (the 'radar depth' of the Cytherean globe amounts to more than 1/50 of a second, and the planetary radius of 6056 ± 1 km has been deduced from its measurement) but reflection from a spinning globe would be also subject to observable shifts in frequency of the echo profiles (due to a different velocity of the individual surface elements along the line of sight). When such echoes are analysed in time and frequency at different positions of the planet in its orbit, it should be possible to ascertain from the observations the period of planetary rotation as well as the orientation of the axis of rotation in space.

Such observations of Venus, conducted over several years at the Jet Propulsion Laboratory in California, disclosed that Venus rotates in a *retrograde* direction (i.e. one opposite to its orbital motion) in a sidereal period of $243 \cdot 1 \pm 0 \cdot 1$ terrestrial days, about an axis inclined by $87^\circ \cdot 8$ to the orbital plane of the planet. Another case of such a retrograde rotation has already been described for Uranus (see Chapter 2) among the major planets; but in the inner precincts of the solar system it is unique. Even more noteworthy is the fact that a period of 243·16 days—virtually identical with that deduced from the radar data—would allow Venus always to show us the same face at each inferior conjunction.

Let us discuss this important result in greater detail to bring out its significance. As we mentioned already at the beginning of this section, the sidereal year on Venus (i.e. the time interval P_V which will elapse until Venus returns to the same position in its orbit around the Sun with respect to the stars) is equal to 224·70 days of our (terrestrial) time, as compared with $P_E = 365 \cdot 26$ days of our own year. Therefore, the period P'_V of the 'synodic* orbit' of Venus relative to the Earth (i.e. the time interval which will elapse before Venus will show the terrestrial observer the same phase again) follows from the relation

$$\frac{1}{P'_V} = \frac{1}{P_V} - \frac{1}{P_E}.$$

Hence $P'_V = 583 \cdot 9$ days—i.e. much longer than the Cytherean 'sidereal year' of 224·7 days, because the Earth (being farther from the Sun) follows Venus along the ecliptic with inferior orbital speed. Accordingly, a little more than one year and seven months will elapse on the average before Venus—escaping us in front—will catch up with us from behind.

On the other hand, the sidereal day $D = 243 \cdot 1$ days on Venus, established recently from the radar data, is defined as a time-interval between two successive meridian transits of the same star as observed from the planet's surface. Therefore, the solar day D' on Venus (corresponding to a time-interval between two successive passages of the Sun through the meridian) should follow from the equation*

$$\frac{1}{D'} = \frac{1}{P_V} + \frac{1}{D},$$

yielding $D' = 116 \cdot 8$ days, such that

$$\frac{P'_V}{D'} = \frac{583 \cdot 9}{116 \cdot 8} = 5.$$

On the Earth, because our planet rotates fast and revolves slowly, a difference between the mean solar and sidereal day amounts to only 3 minutes and 56·6 seconds; the solar day being the longer of the two because the Earth rotates in the same direction as it revolves. However, on Venus, which rotates slowly in the retrograde sense but revolves fast, the solar day is much shorter than the sidereal one, and its duration (corresponding to a sum of the angular velocities of the sidereal rotation and revolution of the planet) is equal to only 116·8 days—exactly one-fifth of the period of Cytherean synodic orbit!

More surprises are in store for us when we inquire about the time-interval between the successive transits of the Earth through the Cytherean meridian. This time interval—let us call it $D_V^{(E)}$—corresponding to a sum of the angular velocities of sidereal rotation of Venus and of the Earth's revolution around the Sun, should follow from the equation

$$\frac{1}{D_V^{(E)}} = \frac{1}{D} + \frac{1}{P_E}$$

which yields $D_V^{(E)} = 146 \cdot 0$ days, such that

$$\frac{P'_V}{D_V^{(E)}} = \frac{583 \cdot 9}{146 \cdot 0} = 4,$$

rendering the time-interval between two successive meridian passages of the same element of the Cytherean surface for the terrestrial observer exactly one-quarter of the Cytherean synodic year!

These striking coincidences imply that, between each two successive conjunctions of the same kind, the Cytherean globe just manages to complete four rotations with respect to the terrestrial observer, but five rotations with respect to the Sun. For an observer situated on Venus (above the clouds), the Sun rises in the west (because Venus rotates in a retrograde direction),

* The right-hand side of this equation consists of a sum rather than difference because the rotation is retrograde.

PLATE 7: Map of Venus in radar illumination (70-cm wavelength) scattered from the planet's surface (*above*), with the corresponding intensity contours (*below*). After Pettengill and others (1970).

reaches the meridian one (terrestrial) month later, and eventually sets in the east—repeating this sequence every 117 terrestrial days, with no significant change of seasons (since the axis of rotation is almost perpendicular to the orbital plane). Thus the Earth gets in opposition with Venus every fifth solar day, when an observer on our sister planet above the clouds would see the same face of full Earth looking down at him exactly at the time of his midnight as a brilliant, dazzling celestial object of −5·6 apparent visual magnitude—much brighter than Venus can ever appear to us. This remarkable resonance strongly suggests a secular influence of *tidal coupling* between the two neighbouring planets; but its mechanism is as yet largely obscure.

A synthesis of the images of Venus in illumination by radar flashes, based on the range-Doppler tracking, is currently making another unique contribution to the study of our sister planet in the form of radar maps of its surface (see Plate 7), providing a first glimpse of what its solid surface looks like underneath the clouds. The maps disclose that the surface of Venus is not as rough as that of the Moon, but it is by no means smooth. In fact, several regions of the planet seem to be mountainous. Whether these are jumbled rock-strewn fields or real mountain chains we cannot as yet say, but we know for sure that underneath her veil the face of the celestial Goddess of Love discloses many wrinkles to the inquisitive radar eye.

5
The Planet Mars

OUR SECOND-NEAREST planetary neighbour is Mars. This planet revolves around the Sun in 686 days and 22·3 hours at a mean distance of 1·524 AU, in an orbit of marked eccentricity ($e = 0{\cdot}093\ 26$) which never brings it closer to the Earth than $55{\cdot}5 \times 10^6$ km. Moreover, this occurs only when Mars happens to be in opposition to the Earth at a time when it passes through its perihelion*, while the Earth is simultaneously in aphelion*. The time-interval between successive oppositions—i.e. the period P'_M, the synodic orbit of the planet—follows again from the equation

$$\frac{1}{P'_M} = \frac{1}{P_M} - \frac{1}{P_E},$$

where P_M and P_E denote the sidereal years on Mars and the Earth. $P'_M = 780$ days and 7 minutes (i.e. the Earth needs approximately 2 years and 50 days to catch up with Mars in space); and successive perihelion oppositions are, on the average, separated by $15\frac{3}{4}$ years.

Even under these conditions the distance of Mars from the Earth is 1·5 times as great as that of Venus at inferior conjunction. A spacecraft sent out to Mars spends, on the average, about seven months on the way—as against a three-month trip to Venus—before it can reach the proximity of its target. This is due to the fact that a spacecraft en route to Mars travels largely against the direction of solar attraction; while for travel towards Venus the opposite is the case.

The size of the Martian orbit renders it the nearest outer planet to us in space, and its proportion is such that the apparent disc of Mars, as seen from the Earth, will not depart from full phase by more than 47°. Mars can appear distinctly gibbous to us at times, but never as a crescent. Its distance from us in space varies between 55·5 and 378 million kilometres, but even at the closest approach the planet does not become brighter than $-2{\cdot}8$ apparent visual magnitude. This fact alone suggests that Mars is not a very large planet; its maximum apparent diameter of $25''{\cdot}1$ at the time of a perihelion opposition corresponds to a mean radius of 3390 ± 10 km, slightly more than half that of our Earth.

The mass of Mars has, in recent years, been inferred with great precision from the observed motion of fly-by spacecraft. The determinations of the mass-ratio Sun : Mars made in 1970 at the Jet Propulsion Laboratory, California Institute of Technology, from the motion of the Mariners have led to the value of

$$\frac{m_{\odot}}{m_{♂}} = 3\,098\,600 \pm 600$$

and

$$Gm_{♂} = 42\,829 \pm 1 \text{ km}^3 \text{ s}^{-2}.$$

Since, for the Earth $Gm_{\oplus} = 398\,601 \pm 1$ km^3 s^{-2}, it follows that the mass of Mars is equal to $0{\cdot}107\,447$ of that of the Earth, or about $6{\cdot}423 \times 10^{26}$ g. In consequence, its mean density proves to be 394 kg m^{-3}, and gravitational acceleration on the surface $3{\cdot}62$ m s^{-2}.

Since the early days of telescopic astronomy the apparent disc of Mars has—unlike Venus—exhibited to the terrestrial observer unmistakable evidence of semi-permanent markings, whose apparent motions disclosed the existence of diurnal rotation not unlike that of the Earth. Extensive observations carried out for more than three centuries have established with great precision that the sidereal day on Mars is equal to 24 hours, 37 minutes, and $22{\cdot}668$ seconds of our terrestrial time, about an axis which is inclined to the planet's orbital plane by $64°{\cdot}8$ (for the Earth, the corresponding value is $66°{\cdot}55$); and the sense of rotation is direct. A sidereal day on Mars is, therefore, only a little more than 41 minutes longer than our own; but the Martian solar day, the time-interval between two successive passages of the Sun through the Martian meridian, is longer than the sidereal day by 2 minutes and 12 seconds (in contrast with a similar difference of 3 minutes and 56 seconds between the solar and sidereal day on the Earth).

Moreover—as on the Earth—the axial rotation of the planet gives rise to centrifugal force which has flattened the Martian globe to a small but measurable extent. The ellipticity ϵ of the meridional section of the planet

$$\epsilon = \frac{1}{190{\cdot}4 \pm 0{\cdot}3}$$

bears a ratio to the centrifugal force which indicates a smaller degree of internal density concentration than is the case for the Earth. In fact, recent investigations of the internal structure of Mars indicate that its central density scarcely exceeds twice its mean density (for the Earth, the corresponding factor is more than 3); and the Martian core, if there is one, cannot exceed 3–5 per cent of the planetary mass (for the Earth it is about one-third). Moreover, the internal pressure scarcely exceeds 10^{11} N m^{-2} even at the centre; the internal temperature should not exceed some 1000–1500 K.

Of semi-permanent markings discerned on the Martian surface by telescopic observations since the seventeenth century, the most conspicuous have been the polar caps—extensive whitish regions, contrasting distinctly with the reddish colour of the rest of the surface, which surround both areographic* poles. Ever since their existence was first noted by Fontana in 1636, their extent has been found to respond to the alternating seasons of the Martian year. The cap surrounding the north pole of the planet extends in winter down to 60°–65° of areographic latitude, and never disappears completely during the Martian summer, while the southern cap disappears for several weeks at the height of the summer. The regularity of this phenomenon obviously suggests that the substance of which the polar caps are composed melts or sublimes with the advent of the Martian spring, to re-solidify again in the autumn. Although the nature of this substance was not safely identified until recently, the phenomenon itself suggests that the Martian globe does not expose to interplanetary space a bare surface, but that it is surrounded—like the Earth or Venus—by a gaseous atmosphere which offers the surface covered by it at least a modicum of physical protection.

The Martian atmosphere

That Mars possesses a gaseous atmosphere is indicated by other observations. Perhaps the simplest of these is the fact that the planet's disc as seen by a distant observer does not appear to be uniformly bright, but is progressively darkened towards the limb*—especially in blue or violet light. In addition, the light of the planet is partly polarized, and the degree of its polarization increases between centre and limb. Such phenomena suggest the presence of an atmosphere whose absorption or scattering dims surface regions illuminated by low sun. In addition, obscuration phenomena have occasionally been observed to impair the visibility of surface details in certain regions of the surface, suggesting interposition of different kinds of clouds which form and dissolve in the course of time. This process would also be impossible without the presence of some kind of an atmosphere.

How dense is the Martian atmosphere, and what does it consist of? The clue to its composition came in 1947, when absorption bands of carbon dioxide were identified in the Martian spectrum by Kuiper and his associates. The actual structure and composition of the Martian atmosphere did not, however, completely come to light until the advent of spacecraft in the 1960s. Six spacecraft have managed to pay close calls on Mars since 1965. The first milestone was the historic flight of U.S. Mariner 4 which, at the time of closest approach, overflew the Martian surface at an altitude of 9840 km. Mariners 6 and 7 of 1969 came within 2000 km of their target; and Mariner 9 of 1971 went into orbit around Mars, while the Russian probes Mars 2 and 3 managed to effect surface landings at about the same time.

A determination of the pressure and density profile of the Martian atmosphere, from the refraction of the spacecrafts' radio signals as they underwent occultation by the Martian limb during their respective fly-bys, disclosed the following facts. The total air pressure above the Martian ground amounts to only 5–6 millibars, less than one-hundredth of a terrestrial atmosphere. This pressure is encountered on Earth about 30 km above the surface. The Martian pressure drops to one millibar at about 25 km, and to one-tenth of a millibar at 50 km. Carbon dioxide represents at least 90 per cent of the Martian atmospheric air mass, and close to 100 per cent if nitrogen and argon (additional constituents possible, but not yet positively identified) prove not to be present in significant amounts. In 1969 carbon monoxide was found by Kaplan, Connes, and Connes to be present in Martian air in a concentration of 0·8 per cent by volume, but the content of water vapour appears to be less than 0·5 parts per thousand by mass, and that of free oxygen still smaller.

The Sun appears less than half as bright from Mars as from the Earth, but the intensity of its light is sufficient to endow Mars with an ionosphere. According to the measurements performed just before the occultation of the Mariner spacecraft by the limb of the planet, the maximum density of free electrons is encountered at an altitude of approximately 120 km above the Martian surface (in contrast to 300 km on the Earth), and amounts to only about 10^{11} electrons per cubic metre (about one-tenth of the maximum density of our ionosphere).

The temperature of Mars

Another feat performed by the Mariners was measurements of Martian temperatures both on the ground and aloft. Measurements of the thermal radiation emitted by the planetary surface in the infrared and microwave regions of the spectrum had long shown that temperatures on the Martian surface are much lower than on the Earth. Measurements of the intensity of infrared radiation of the planet made through the 8–12 μm window of our atmospheric transparency by Sinton and Strong (1960) and Pettit (1961) indicated that a temperature as high as 300 K prevails at the sub-solar point in the Martian tropics; this drops to some 250 K towards sunset. At the rim of the polar cap temperatures lower than 210 K were registered.

Night temperatures on Mars (or any other outer planet) cannot, of course, be measured directly, because little of the night hemisphere can be visible to us. However, at least an indication of their level can be obtained from the measurements of the thermal emission of the Martian disc in the domain of radio-frequencies (at wavelengths of 0·1–10 cm) to which our atmosphere is transparent. Repeated measurements by a number of investigators have disclosed that, at wavelengths longer than one centimetre, which originate at a depth of several times the wavelength, the measured intensity of thermal radiation corresponds to a temperature of 200–205 K which remains constant day and night. The actual temperatures prevailing on the exposed surface depend, of course, on the thermal inertia of insulating surface layers, and with such estimates of these as can be made it would seem that the night temperatures on Mars should be expected to sink considerably below $-100°$C.

The historic missions of Mariners 6 and 7 in 1969, which also performed the measurements of Martian thermal emission at the time of their closest approach to the surface of the planet, gave us one most interesting piece of information: namely, that on the rim of the north polar cap the ground temperature proved to be only 155 ± 10 K. If we compare this value with the temperature of 148 K at which carbon dioxide solidifies at a pressure of 5–6 millibars, the message of the measurement become unmistakable: the Martian polar caps are not icefields of frozen water like those in the polar regions of our Earth, but rather frozen layers of ‘dry ice’, or solidified carbon dioxide. This fits in with what we know now of the composition of the Martian atmosphere, with its preponderance of carbon dioxide and virtual absence of water, and so one of the age-long problems of the Martian environment appears to have been solved.

With the approach of the winter season, the arctic air temperatures on Mars apparently become low enough for some carbon dioxide to freeze out of the atmosphere and cover the ground with snowshowers of dry ice, which melts again (partly in the north, more completely in the south) with the advent of spring. Such snowcover appears, even at the height of the winter, to be quite thin—from inches to yards in depth. No permanent glaciers are formed: a far cry indeed from the massive icecaps covering the polar regions of our own planet. As the solid carbon dioxide never melts but sublimes directly into the air, no liquid ever flows from the Martian polar regions to the tropical belt, and there is no need for any 'canals' to drain it off.

It should be added, of course, that all the temperatures we have mentioned so far refer to the solid surface, or the base of the Martian atmosphere. Appropriate computations disclose that, at an altitude of no more than 20 km above the surface, the mean daytime temperature should drop to some 200 K; and at 50 km to less than 120 K. Another 50 or 70 km up—in the Martian ionosphere—the prevalent temperature may exceed 470 K, but this is much lower than the temperatures in excess of 1300 K found in the denser parts of the corresponding terrestrial layers.

On the other hand, the smaller gravitational acceleration on Mars will cause the density and pressure of Martian air to diminish more slowly with increasing altitude than is the case on Earth. While the Martian air is at no level actually denser than our own terrestrial atmosphere, the surface disparity in their density and pressure generally diminishes with height, and the same is true of the Martian ionosphere as well.

The Martian surface

Let us descend from the atmosphere again to the Martian surface, and return to the markings discernible on it. As long as the only observational instruments available were optical telescopes, two classes of such markings were recorded and mapped (see Plate 8). One comprised the 'dark markings' of lower reflectivity, covering less than one-third of the visible surface, generally regarded as 'continents', and sometimes interconnected by 'oases' or 'canals' on the background of vast expanses of brighter areas of reddish colour, generally regarded as 'deserts'. Unlike the polar caps, the general outlines of the bright and dark markings have remained sensibly stable for decades if not centuries; but for the 'oases' or 'canals' this is much less certain. In fact, topographic details of this type reported by different visual observers were often at variance with each other, and so close to the limits of the optical resolution of their telescopes that doubts were repeatedly expressed about the reality of such phenomena. As we have found out in the past few years, such doubts were indeed well-founded.

The first contribution of space astronomy to the study of Mars was the establishment of radar contact with this planet in 1963. This was considerably more difficult than establishing radar contact with Venus, not only because of the greater distance (unlike light intensity, which diminishes with

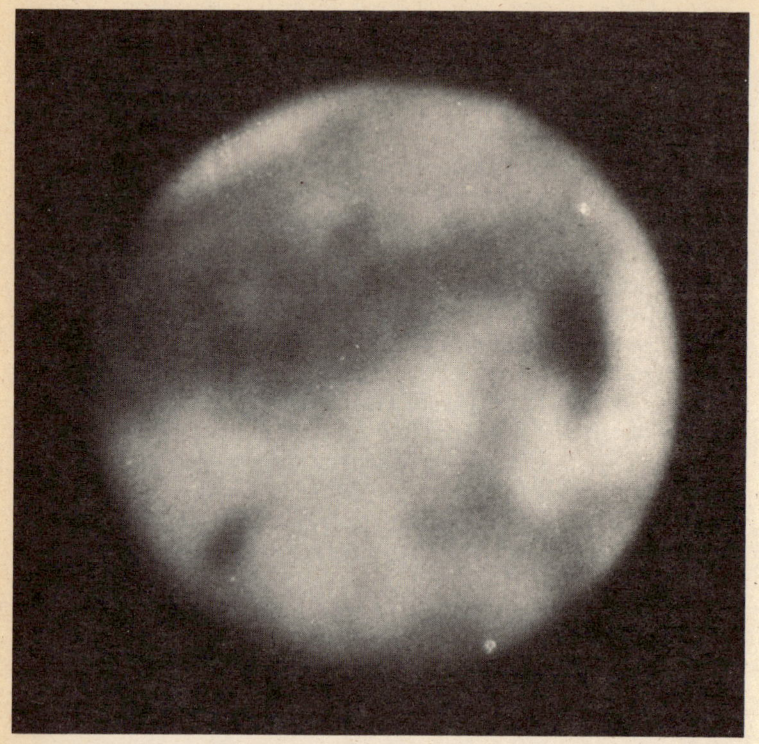

PLATE 8: Photograph of the planet Mars, taken by H. Camichel with the 15-inch refractor of the Observatoire du Pic-du-Midi.

the square of the distance, the strength of a radar echo—because of reflection—attenuates with the inverse fourth power of the distance of the reflecting screen), but also because a much higher rate of planetary spin correspondingly widens the frequency profile of the returning echoes.

However, such results as were obtained in spite of these difficulties disclosed that the structure of the Martian surface is quite rough—more so than appears to be the case for Venus. The strength of the returning echoes is consistent with a Martian surface composed mainly of dry rocks. Limonite, an oxide of iron, appears to be a suitable candidate for predominance from its colour as well as its dielectric properties. It is now certain that no appreciable part of the Martian surface is covered with water.

But even the large-scale structure of the Martian surface appears to be warped to an unexpected extent. Different methods of analysis of the radar echoes lead to consistent results demonstrating that the surface exhibits differences in elevation as great as 12–16 km over long gradual slopes covering wide areas. Actually, there does not seem to be much level ground on Mars and, surprisingly, no obvious correlation seems to exist between the different types of markings visible on the surface and their elevation—high and low levels occur equally in the 'bright' and 'dark' areas. To give an example, Syrtis Major, the largest and most conspicuous dark marking on the Martian globe (visible through relatively small telescopes) proved to be a giant slope rising by more than 10 km in elevation from its (astronomical) western to eastern edge; comparable differences exist elsewhere as well.

It is true to say that if the oceans were siphoned off the terrestrial surface, and if the latter were scanned by a radar beam from outer space, level-differences larger than 10–12 km would be established between the abyssal plains of ocean floors and the high continental plateaus of central Asia. However, the Earth is twice as large as Mars; a level-difference of 10–12 km on Mars would proportionally correspond to a terrestrial difference of 20–25 km, which exceeds the difference in elevation between the top of Mt. Everest and the depth of the Marianna Trench in the Pacific. We are led to the inescapable conclusion that the surface of Mars has been levelled off by erosion less than that of our Earth, and that the role of air and water—the principal levelling agents—has been much smaller on Mars not only now, but throughout its entire astronomical past.

That this is indeed so is attested not only by a greater proportional large-scale warping of the Martian surface, but also by its smaller-scale structure as disclosed by the views of its landscape televised to us in recent years by Mariners 4, 6, and 7. Until the advent of spacecraft, limitations of astronomical optics as well as atmospheric conditions prevented telescopic resolution from the Earth of details smaller than some 300 km in size on the Martian surface. The Mars-bound Mariners of the 1960s enabled us to increase this resolution by a factor of the order of 100, and more than 1000 in 1971. In doing so, they opened up for us a window into another world.

What did the views televised by the Mariners disclose? A stark, arid landscape (see Plate 9), very mountainous in places, and profusely dotted with pockmarks identical in size, shape, and other general characteristics to the craters so familiar to us on the Moon. As on the Moon, some are hundreds of kilometres in diameter; their walls are 3–4 km deep; many also possess 'central mountains' with which we are familiar from lunar topography.

Their origin—or at least the origin of most of such formations—is

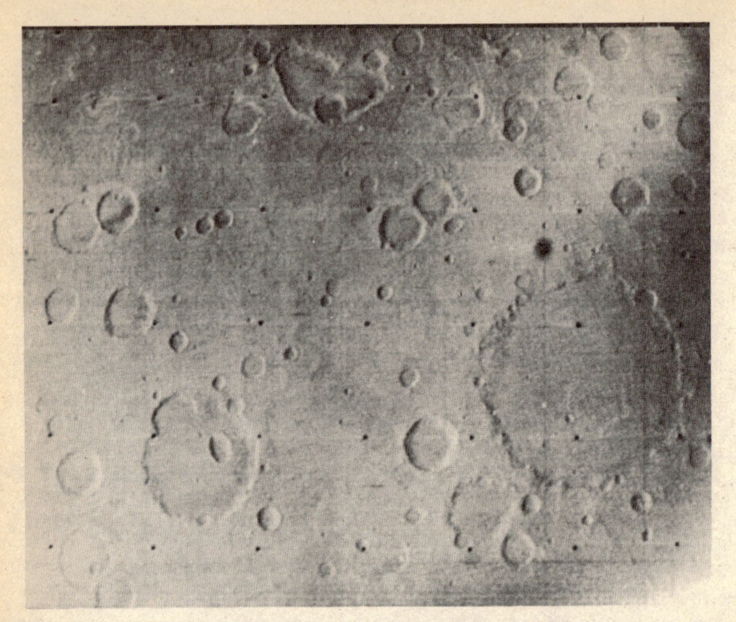

PLATE 9: Martian landscape at a close range, televised by the Mariner 6 spacecraft at the time of its close approach to the planet on 30 July 1969. The surface area seen on the frame is about 700 × 900 km in size, and the largest crater to the right is 260 km across.

likewise in no doubt: they represent impact craters of the type we shall discuss in Chapter 7 in connection with our Moon. The surface of Mars (unprotected, like the Moon, by any atmosphere of appreciable density) represents just another cumulative scoreboard of the celestial target-practice of meteors and meteorites, asteroids and comets, and because the Martian orbit just skirts the inner border of the ring of asteroids (Chapter 8), it may in the past have suffered many more hits by them than our Moon, which lies in the more protected inner precincts of the solar system.

The heavily pockmarked face of Mars bears witness to the great age of its surface, and testifies to a well-nigh complete absence of any erosion processes that could have levelled off this surface relief after long intervals of time.

One of the most telling results supplied by the Mariners has been a virtual disproof of the existence of 'canals' on Mars, which had so exercised the human imagination in the past hundred years; the spacecraft found that they did not exist. The illusion of their existence must have been due to a combination of actual surface roughness and shadows cast by its relief, the details of which are below the limit of optical resolution of terrestrial telescopes. The disappearance of the 'canals' under the more searching glance of the spacecraft eyes should caution over-optimistic observers from fishing for information inside the limit of resolution of their instruments.

Just as the spacecraft showed there are no canals on Mars, as so many people used to imagine, they also failed to discover there any mountain chains which could—as on the Earth—be due to the folding of the crust. Indeed, the Martian crust, with its ability to support more than 10 km-altitude differences, and the absence of folded mountains, gives every impression of possessing rigidity far greater than that of the terrestrial crust. This rigidity is probably the consequence of a relatively low temperature in the interior. Since Mars is but one half of the Earth in size, and contains only a little more than ten per cent of its mass, it could not have developed and stored nearly as much internal heat as our own planet. All external manifestations of its surface known to us so far appear to be consistent with this view.

Furthermore we are practically certain that, unlike our Earth, the planet Mars contains in its interior no metallic core of any size. We conclude this not only from the absence of any pronounced concentration of density in its interior (which would be consistent with the observed polar flattening of the planetary globe), but also from the fact that—unlike the Earth, but like Venus—Mars does not possess any measurable magnetic field. Magnetometers aboard the Mariners failed to detect Martian magnetic moments more than one or two ten-thousandths of that of the Earth; and for a planet rotating almost as fast as our Earth so complete an absence of a magnetic field signifies that the metallic 'dynamo' is not a part of the equipment stored in the Martian interior. Thus, like Venus, Mars is not surrounded by any van Allen belts of charged particles possessed by our Earth, though it does possess an ionosphere.

Under all these circumstances, could a planet like Mars ever have given rise to life? Whatever hopes we may have entertained in this respect before the advent of spacecraft, the results of their first three space missions in 1965 and 1969 were enough to shatter these hopes beyond repair. The Martian environment appears to be almost as inhospitable—if not hostile—to life as our Moon; and should we, against all our present expectations, find any traces of living matter on its surface, it would be of interest to the microbiologists rather than to the humanists in our midst. The hope of finding life on Mars has evaporated, together with the canals, into the thin Martian air. It is now almost certain that as living beings we are alone in the solar system.

Satellites

There is one more fact to be noted before we conclude this brief survey of our present knowledge of the physics and astronomy of Mars. The planet is attended by two satellites, Phobos and Deimos, discovered by Asaph Hall in 1877 (see Plate 10). These satellites are indeed most extraordinary astronomical bodies. Phobos, the inner satellite, revolves around its central planet at a mean distance of 9354 km (1·378 times the planet's equatorial radius) in a period of 7 hours and 39 minutes—i.e. much faster than the planet rotates about its axis—while Deimos revolves around Mars at a distance of 23 490 km (3·460 times the Martian radius) in 30 hours and 17 minutes.

By the rapidity of its motion across the Martian sky, Phobos almost resembles an artificial satellite of the Earth. The period of its synodic orbit (after which it returns to the same phase) is only twelve seconds longer than the sidereal month. Since, however, its orbit is prograde* and its period is shorter than that of the Martian axial rotation, for an observer on the Martian surface it will move from west to east in the sky but will not remain above the horizon of any place longer than 3 hours and 10 minutes, in the course of which it will run through a major part of its phase cycle. Moreover, the inclination of its orbit to the ecliptic is such that (for a suitably situated observer on the Martian surface) Phobos can transit across the disc of the Sun, whose apparent diameter at the distance of Mars is only 21 minutes of arc. However, so rapid is the motion of the satellite that even a central transit across the Sun would last less than 30 seconds.

The orbit of the satellite Deimos around the planet Mars is likewise prograde; but as its period exceeds the Martian sidereal day by only 5 hours and 40 minutes, this satellite moves slowly (by about 2°·8 per hour) from east to west on the Martian sky, and passes through the meridian only after a time interval of 132 hours while the planet has rotated 5·36 times about its axis. As, moreover, the 'synodic month' of Deimos differs from its sidereal period of revolution by only 3·3 minutes, it follows that between successive passages through the meridian the satellite should run through its phase cycle more than four times!

The absolute dimensions of Phobos were recently determined from an image of this satellite televised in November 1971 by Mariner 9 (Plate 10). Phobos turned out to be an elongated body, about 25 × 21 km in size, and its reflectivity proved to be as low as that of any other body encountered in the solar system so far: only about six per cent of visible sunlight incident

PLATE 10: Above: satellites of Mars—Phobos and Deimos—photographed by G. P. Kuiper with the 82-inch telescope of the McDonald Observatory. Below: a close-up view of the Martian satellite Phobos, televised by the U.S. Mariner 9 spacecraft at the end of November 1971 from a distance of 14 683 km from the target. Phobos is an elongated body, some 25 by 21 km in size; and its surface bears ample evidence of cratering caused by external impacts. This is what a typical asteroid would look like if we could approach it at a similar distance, and there is little room for doubt that both Martian satellites—Phobos and Deimos—as well as several outer satellites of Jupiter—are nothing else but captured asteroids.

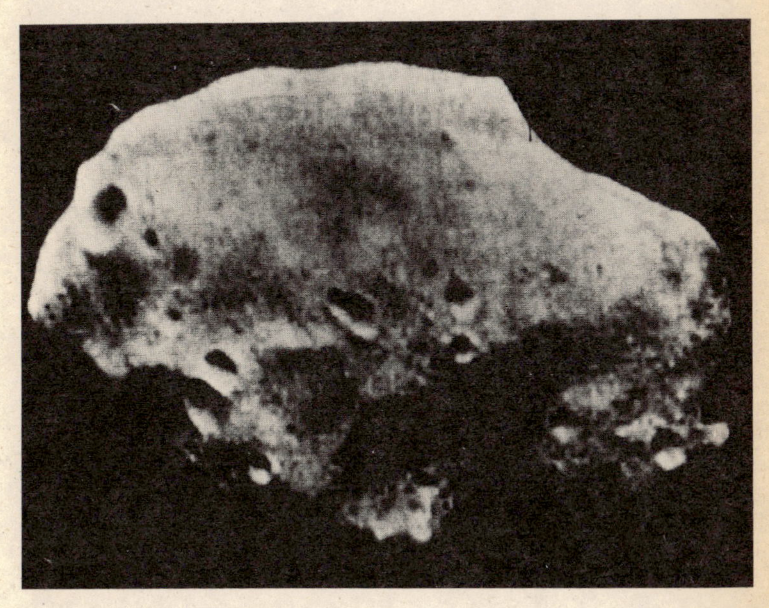

upon its surface get scattered by it in all directions. Deimos—a much fainter object—is obviously smaller, probably no larger than 6 to 8 km across. It is a mere chip of a rock moving about rather lazily in space, so small that an observer on the Martian surface could not discern its angular size with the naked eye.

6 Mercury and Pluto

THE TWO REMAINING planets of the terrestrial group, Mercury and Pluto, occupy positions of special significance in our solar system as the innermost and outermost planets of the system. Mercury is in scorching proximity to the Sun; Pluto is at the outer rim where the Sun appears only as a star in the daytime sky, and its light barely interrupts the monotony of the cold and darkness of interstellar space.

Because of the difficulty of observing astronomical bodies stationed at such extreme outposts, our knowledge of the physics and astronomy of Mercury and Pluto is still far more limited than is the case with our closer neighbours. Mercury revolves so close around the Sun that it never elongates from it for us by more than 28°, and this makes observation difficult; while Pluto is so far away that a fairly large telescope is needed to make it visible from the Earth at all.

Both these planets revolve around the Sun in orbits that are markedly eccentric ($e = 0{\cdot}2056$ for Mercury and 0·249 for Pluto), and inclined much more to the ecliptic (by 7°·0 for Mercury and 17°·1 for Pluto) than for any other planet. However, whereas the actual distance of Mercury from the Sun oscillates between 46 and 69 million km (its mean value being 57·0 million km or 0·387 AU, a distance which light traverses in 3·22 minutes), the mean distance of Pluto is 5910 million km, or 39·52 AU: light from the Sun needs 328 minutes to reach this outermost of all known planets.

The orbital periods of Mercury and Pluto differ accordingly, for while the sidereal year on Mercury is only 87·969 days, Pluto needs a little more than 349·17 years to accomplish one complete orbit around the Sun. The synodic year on Mercury—i.e. the time-interval after which Mercury (an inner planet) will exhibit the same phase to the terrestrial observer—is, however, equal to 115 days 21 hours of our own time; and the actual distance of Mercury from us fluctuates from interior to superior conjunction, between 79 and 218 million km. During this time, the apparent brightness of the planet oscillates between $-1 \cdot 2$ and $+1 \cdot 1$ apparent visual magnitude (i.e. from almost the apparent brightness of Sirius to that of Aldebaran); while the apparent brightness of Pluto is almost constant and is $14^{m} \cdot 6$ (or $15^{m} \cdot 3$ in photographic light).

One other thing that these two planets have in common is their relatively small size and mass. On account of Mercury's proximity to the Sun, it is difficult to carry out accurate measurements of the angular diameter of its apparent disc (never greater than 13″) in twilight sky low above the horizon, and the measurements of its apparent size during its occasional transits across the Sun (a few times per century) are subject to peculiar errors of their own. Fortunately, however, Mercury is close enough to us to be contacted by radar; and an analysis of its radar echoes, obtained repeatedly since 1962, shows the diameter of this planet to be 4879 ± 1 km, or 0·3829 times the mean diameter of the Earth. Pluto, on the other hand, is hopelessly out of range for any radar contact, and an angular diameter of its apparent disc of $0'' \cdot 13$, measured in 1965, is still subject to considerable uncertainty; but corresponds to a diameter between 6000 and 7000 km.

The mass of Mercury is known to be 1 part in 6 000 000 of that of the Sun (with an uncertainty of 1–2 per cent) or 0·053 of that of the Earth. A combination of this mass with Mercury's radius of 2440 km leads to a mean density close to 5500 kg m^{-3} for the planetary globe—about the same as for our Earth, but unexpectedly high for a planet of as small a mass as Mercury's. The same appears to be true of Pluto. Its mass is the least well known of all the planets, but seems to be close to eighteen per cent of that of our Earth and if this is so, the mean density of the Plutonian globe is unlikely to be less than that of Mercury, and may be higher.

Radar observations of Mercury have disclosed that this planet rotates about an axis almost perpendicular to its orbital plane in a (sidereal) period of 58·6 days of our own time; the sense of rotation is direct. A sidereal day of $58^{d} \cdot 6$ of terrestrial time on Mercury corresponds, however, to a solar day of 176 days of our own. This latter period is just three times the length of the Mercurian sidereal day (and two-thirds of its sidereal year). Thus, contrary to some previous belief, Mercury does not always show the Sun the same face. To an observer on one surface of the planet the Sun, a huge fiery disc of apparent diameter 2–3 times as large as we see on the Earth would rise in the east and set in the west every 176 terrestrial days. At the time of each perihelion passage the Sun would appear to pause in the sky and actually reverse the direction of its motion for about two weeks. This should cause excessive heating at the sub-solar point at that time, giving rise to two super-tropical regions situated on the planet's equator opposite to each other. Seasonal effects are probably non-existent—as on Venus—because of the negligible inclination of the planet's equator to its orbital plane.

Pluto seems likewise to rotate about an axis (of unknown orientation) in a much shorter period of 6 days, 9 hours, 17 minutes, as evidenced by the measured small changes of its brightness fluctuating in this period. This is due no doubt to successive passages through the meridian of regions which reflect sunlight to different extents. Pluto is too far away, and its disc is too small, for us to see such regions as bright and dark spots, but photometers have disclosed their alternation to us even though the spots themselves are below the limit of the resolving power of our telescopes.

The reflectivity of the Mercurian surface for centimetre waves of radar signals has proved to be very similar to that of our Moon. We surmise from this that the physical structure and chemical composition of the Mercurian surface—likewise unprotected by any detectable atmosphere—may be similar to that found on the Moon; in particular, this surface may be covered with the same kind of strong regolith. With sufficiently powerful telescopes we should no doubt find the surface of Mercury to be pock-marked with impact craters like the lunar or the Martian surface.

The long duration of the solar day on Mercury leads us to expect that the extremes in temperature on its surface in daytime and at night are likely to be large. From measurements of the intensity of its thermal emission in the infrared part of the spectrum, the temperature at the sub-polar point of the Mercurian surface was found (in 1970) to be as high as 620 K as a result of its close proximity to the Sun. At night, Mercury becomes almost as cold as the Moon. The prevalent temperature is then close to 110 K, in spite of Mercury's proximity to the Sun, because nights on Mercury are $6\frac{1}{2}$ times as long as on the Moon. A difference in temperature of more than 500 K between day and night on the same spot entails a thermal inertia of the surface material comparable with that on the Moon, and consistent with the relative weakness of the radar echoes returned by the Mercurian surface.

Nothing whatever is known so far of the structure of the surface of Pluto, except that its reflectivity is not everywhere the same; as is mentioned above its apparent brightness is known to fluctuate somewhat as regions of unequal surface brightness pass periodically through the meridian. At its mean distance of 39·5 AU from the Sun, sunlight is already so diluted that its absorption could not warm up the surface to more than 50 K even in daytime; but how close the actual temperatures come to this limit has yet to be established.

7
Our Moon

WE BEGAN THIS description of the principal properties of the terrestrial planets with an account of our Earth as the largest and best-known representative of this class of astronomical bodies: we shall end it with a similar account of our Moon. It is entirely appropriate that this be so; for not only does the Moon represent the extreme end of the mass range of the terrestrial planets—possessing a mass not much larger than the minimum necessary for a solid body to assume spherical form—but, quite recently, the Moon has also become the first celestial body on which man has set foot.

The results of exploration of the lunar surface by means of spacecraft in the 1960s, superimposed on previous decades and centuries of telescopic work, have provided us with a more complete picture of the Moon and its environment than we possess for any other celestial body except the Earth. The aim of the present chapter is to give a brief summary of our present knowledge of the Moon and place it in proper perspective with respect to the other bodies constituting the solar system.

The fundamental facts and figures about the motion and physical attributes of the Moon are well known. The orbit of the Moon around the Earth is approximately an ellipse (distorted appreciably by the attraction of the Sun or, more precisely, by the difference of the solar attraction on the Moon and the Earth), inclined to the orbital plane of the Earth around the Sun by $5^\circ 8' 43''\cdot 4$. The semi-major axis of this ellipse (the mean distance between the Earth and the Moon) was established within a few per cent of the correct value from the relative duration of different phases of lunar eclipses by Hipparchos in the second century B.C. In more recent centuries, it has been measured by celestial triangulation (using the known dimensions of the Earth as a baseline), and in the last few years by the measured time-lag of radar and laser echoes returned from the lunar surface. This distance (384 402 km) and the distance between the Earth and the Moon fluctuates between 356 and 407 thousand kilometres on account of a mean orbital eccentricity $e = 0{\cdot}0549$. The mean distance is equal to 60·27 times the Earth's equatorial radius, or 0·002 57 AU. It represents, therefore, less than one per cent of the distance separating us from our two nearest celestial neighbours, the planets Venus and Mars, even at the times of their closest approach. Light traverses this distance in 1·28 seconds and an average spacecraft from the Earth can reach the Moon after a free flight of 65–70 hours.

Moon in the sky: months and eclipses

The mean period of revolution of the Moon around the Earth is equal to 27·321 661 50 . . . mean solar days, or 27 days, 7 hours, 43 minutes, and $11\frac{1}{2}$ seconds, after which interval the Moon returns to the same place in the sky. This is the so-called sidereal month. During this time the Sun has, however, moved eastward by approximately one-twelfth of the entire circle; and, consequently, after the lapse of one sidereal month the Moon has not returned to the position in which it would show us the same phase in solar illumination. The time-interval between two successive identical phases of the Moon (the synodic month) is, therefore, longer than the sidereal one: $29^{d}{\cdot}530\,588\,3$ . . . or 29 days, 12 hours, 44 minutes, and 2·8 seconds. However, if we define a month as the time after which the Moon will return to the same relative place in its orbit around the Earth, this so-called anomalistic month will likewise be longer than the sidereal one (because of the secular advance of the line joining the positions of the perigee* and apogee*—the apsidal line of the lunar elliptical orbit. It is $27^{d}{\cdot}554\,550\,3$, or 27 days, 13 hours, 18 minutes, and 37·4 seconds. Lastly, a time-interval between two successive transits of the Moon through the nodes* (i.e. the points at which the lunar orbit intersects the ecliptic)—the so-called draconic month—is shortest of them all (because the nodes recede) and is $27^{d}{\cdot}212\,220$ or 27 days, 5 hours, 5 minutes, and 35·8 seconds.

The draconic month is of considerable significance in connection with the recurrence of solar or lunar eclipses. As is well known, such eclipses can occur if the Moon finds itself between the Sun and the Earth, and eclipses the Sun, partially or totally, for the relevant part of the terrestrial surface. If the Moon enters the shadow-cone cast by the Earth into space in solar illumination, there is an eclipse of the Moon. In either case the occurrence of eclipses depends critically on the relative positions of lunar nodes on the ecliptic. If the synodic and draconic months were identical, each new and full Moon would occur in the same relative position with respect to the nodes; therefore, we should either have a solar eclipse at each new Moon (and a lunar one at each full Moon), or none at all.

We know very well that this is not the case—a fact which, in itself, is sufficient to prove that the positions at which the lunar orbit intersects the ecliptic do not remain fixed in space. However, it happens (and the reader can easily verify this) that 223 synodic months are almost equal to 242 draconic months, the difference between the two multiples being only 51 minutes 41·2 seconds; so that the positions of the Moon relative to the nodes of its orbit should be almost the same every 6585 days—or 18 years and 10–11 days (depending on whether this interval contains 4 or 5 leap years). In consequence, should an eclipse of the Sun or the Moon occur at a certain time, it should recur at the same place after 6585 days, a period already known to the Chaldeans twenty-four centuries ago, and described by the Greeks as *Saros*. Even closer is the coincidence between 716 synodic and 777 draconic months, which leaves a discrepancy of only 9 minutes 46·1 seconds. Eclipses should therefore recur more closely after 21 144 days or just under 58 years, and there are still longer intervals for which this is even more accurately the case.

But let us return to the Moon in the sky, as we see it as an apparent disc of an angular diameter just over half a degree—or, more accurately, $1865''\cdot 2$, oscillating by $204''\cdot 8$ between perigee and apogee—which at a mean (geocentric) distance of the Moon corresponds to a mean radius of the lunar globe equal to 1738 km. The Moon is essentially a sphere of this radius, and departures of its actual surface from this sphere do not generally exceed ± 2 km, and are geometrically quite complicated. The Moon is, therefore, a little more than one-quarter of the size of the Earth. Its surface covers an area of just under 38 million km^2, and its volume amounts to $2\cdot 2 \times 10^{10}$ km^3: very approximately 2 per cent of that of the Earth.

This entire surface of the Moon is not visible from the Earth, because of synchronism between its axial rotation and its orbital revolution. The revolution of the Moon around the Earth—and, with the Earth, around the Sun—are not the only motions performed by our satellite. It also rotates, with a uniform angular velocity, about an axis fixed in space so that the lunar equator is inclined by $1^\circ 32'$ to the ecliptic and by $6^\circ 40'$ to the Moon's orbital plane. It rotates in exactly the same period as it revolves, thus showing us almost exactly the same face each month. Not quite exactly so, however, for several reasons. The first is the fact that although the Moon's axial rotation is uniform, the angular velocity of its revolution in an elliptic orbit (varying as it does with the inverse square of the radius-vector) is not, being sometimes ahead, and at other times behind, the orbital motion. This difference can cause angular displacements in longitude of positions on the lunar surface by as much as $7°54'$ as seen from the Moon's centre—a phenomenon known as the lunar optical libration in longitude.

Secondly, the lunar axis of rotation is not perpendicular to the orbital plane: its inclination deviates from $90°$ by $6°40'$. We can therefore see sometimes more of one lunar polar cap, and at other times more of the other, in the course of each month. This phenomenon gives rise to an optical libration in latitude, by $\pm 6°41'$. Again, when the Moon is rising for the observer on the surface of the Earth, we look a little over its upper edge and see more of that part of the Moon than we should from the centre of the Earth; when the Moon is setting the converse is true. This diurnal libration (not of the Moon, but of the terrestrial observer) can attain $57'\cdot 1$ (i.e. the angular diameter of the Earth as seen from the Moon) and superposes on all other librations, optical as well as physical, to enable us to see from the Earth considerably more than one-half of the lunar surface. Altogether, 59 per cent of the entire lunar globe can be seen from the Earth, and only 41 per cent is permanently invisible; 18 per cent is alternately visible and invisible.

In October 1959 the cameras aboard the Russian Luna 3 unveiled for the first time the principal features of a major part of the lunar far side. About 13 per cent of it remained uncharted until July 1965 when the Russian Zond 3 succeeded in recording all but a small fraction of the remainder. Between 1966 and 1968 the entire front of the Moon as well as the far side was re-photographed with superior resolution by the five U.S. Lunar Orbiters. Thanks to their work we are now in possession of almost as complete a coverage of the entire surface of the Moon as we have for our Earth.

Physical properties of the Moon

The *mass* of the Moon is obtained from a determination of the Earth : Moon mass-ratio, which has in recent years been determined with great accuracy (from the observed acceleration of spacecraft impinging on the Moon) to be $81\cdot 302 \pm 0\cdot 001$. Since (see Chapter 3) the mass of the Earth is equal to $5\cdot 9 \times 10^{24}$ kg, the mass of the Moon must be $7\cdot 35 \times 10^{22}$ kg. A mass of this order of magnitude may seem large in comparison with familiar terrestrial standards but on the cosmic scale it is only a tiny speck.

The mean density of the lunar globe is not at all unusual; for dividing the mass just found by the lunar volume of $2\cdot 20 \times 10^{19}$ m^3, we find its mean density to be $\rho_m = 3340$ kg m^{-3}—only a little higher than the density of common granitic rocks of the Earth's crust (2780 kg m^{-3}), and considerably less than the mean density of the terrestrial globe (5520 kg m^{-3}). The gravitational acceleration $Gm_{☾}r^{-2}$ on the lunar surface is, therefore, equal to only 1·62 m s^{-2}, and the velocity of escape $(2Gm_{☾}r^{-1})^{½}$ from the lunar gravitational field is close to 2·38 km s^{-1} (in contrast with its terrestrial value of 11·2 km s^{-1}).

The relative smallness of the mass of the Moon, and the low velocity of escape from its gravitational field, entail several further consequences; and perhaps the most important one is the complete *absence of any atmosphere* which could shield its surface from direct contact with outer space. By the same token, the Moon cannot maintain any liquid on its surface for any significant length of time. Its surface must, therefore, be regarded as bone-dry and must always have been so. No feature extant today could have been formed, or modified, by the effects of running water, or by repeated freezing and melting of water entrapped in surface rocks. Thus one of the most important agents responsible for shaping the Earth's surface has been completely absent from the Moon, and cannot be invoked to explain any structural characteristics of its surface. Likewise, we cannot expect to find on the Moon any rocks which could have originated by sedimentation, and the fact that none of the missions which have soft-landed on the Moon has found any should cause no surprise.

The internal structure of so small a cosmic globe as the Moon can be expected to be simpler, and less differentiated, than is true for the Earth or other larger terrestrial planets. Simple facts in our possession suggest that this is indeed the case. The average density of surface rocks brought back to the Earth from the Moon by the Apollo 11–15 missions ranges between 3100–3500 kg m^{-3}—considerably greater than that of the granitic rocks of the terrestrial rust (2800 kg m^{-3}), no doubt because of the higher admixture of metals (titanium, iron) that are present in them. More significantly, the density of lunar surface rocks appears to be virtually the same as (or only slightly lower than) the mean density (3340 kg m^{-3}) of the lunar globe as a whole. This ratio must be even closer to unity if one considers that the mean density of 3340 kg m^{-3} refers to self-compressed material, while the measured densities of surface rocks refer to zero pressure. The fact that this ratio is so close to unity for lunar material seems to preclude any large differentiation of the material in the interior (or at least of differentiation that would be accompanied by an appreciable change in density as well). All dynamical indications of the internal structure of the Moon (based on the motion of the Moon in space or physical librations about its centre of gravity) point to its being an essentially homogeneous body.

If this is the case it should be easy to estimate the pressure that should prevail inside a self-gravitating homogeneous configuration of lunar mass. The central pressure P_c should, in fact, be given by the simple formula

$$P_c = \tfrac{2}{3}\pi G \rho_m^2 r^2,$$

where the gravitation constant $G = 6{\cdot}668 \times 10^{-11}$ m^3/kg s^2. If we insert in the foregoing expression the mean density $\rho_m = 3340$ kg m^{-3} and the radius $r = 1{\cdot}738 \times 10^6$ m, we find the corresponding value of central pressure, P_c to be $4{\cdot}71 \times 10^9$ N m^{-2}. Such pressures are exceeded at a depth of only 150 km in the terrestrial mantle, and can readily be attained in the laboratory. Changes in density exhibited by common silicate rocks under such compressions have already been measured; and on the basis of available petrological evidence we conclude that if the Moon consists of rocks similar to the terrestrial silicates—as its mean density seems to suggest—a homogeneous globe of silicate rocks of the lunar size and a mean density of 3340 kg m^{-3}, should, at zero pressure, possess a density of only 3280 kg m^{-3}, which the central pressure of 47 kilobars should increase to 3410 kg m^{-3}. Nothing that we know so far conflicts with this hypothesis.

The pressure inside the Moon should, therefore, be of the order of 10 kilobars throughout most part of its mass. This is about 10 times the crushing strength of granite and other similar rocks; their structure should therefore be expected to give in to hydrostatic pressure, and the entire material should behave as a fluid. Given a sufficiently long time (no doubt short in comparison with the age of the Moon), material which on a shorter time-scale would behave as a solid is bound to get crushed under its own weight to settle down to a form of minimum potential energy—a sphere. This is why not only gaseous stars like the Sun, but also solid bodies like the major or terrestrial planets, are spherical. The maintenance of non-spherical shape remains a prerogative of celestial bodies such as very small satellites, asteroids, or meteorites, whose mass is too small to give rise to internal pressures capable of overcoming the molecular forces of the solid state.

That the Moon—or at least its crust—must possess considerable rigidity is shown by the extent to which it can resist a tendency towards isostasy*, and support deviations from it which are much larger than those found on the Earth (when both are scaled to the same size). Perhaps the most interesting aspect of this was brought to light by the discovery from their effects on the motion of overflying artificial satellites of relatively large but localized positive gravitational anomalies on the Moon. The observed accelerations of the motions of the satellites can be produced only by local mass concentrations (mascons for short) of which there is no visible evidence on the surface. Moreover, the observed rates of change in orbital motion indicate that the mascons responsible for them must be relatively small in size (50–200 km) and located at a shallow depth (50–100 km) below the surface. Their masses are between 50 and 100 millionths of that of the whole Moon, and they exert pressures of 1–10 kg cm^{-2} on their substrate.

Muller and Sjögren, who discovered these mascons in 1968 from an analysis of the perturbations of U.S. lunar orbiting satellites, found that most of their positions corresponded pretty well with the so-called circular maria on the lunar surface. Suppose (and we shall return to this idea later) that these mascons represent leftovers of asteroidal bodies whose impacts created the circular maria—i.e. cosmic bullets that hit the Moon in very distant past and became imbedded in its crust at a shallow depth below the surface. How long could they remain perched there, defying the persistent efforts of gravity to pull them down towards the centre? A simple analysis discloses that if the Moon possessed the same degree of rigidity as the terrestrial mantle, its mascons would be bound to sink to depths at which orbiting satellites could no longer sense their presence in a time of the order of 10^7 years—within less than 1 per cent of the probable age of the Moon.

This explains, incidentally, why there are no similar mascons on the Earth (which should have absorbed a comparable number of cosmic impacts, per unit area) at the present time; they could linger near the surface for only a relatively short time. Since, however, they are very much present on the Moon, the only plausible inference we can draw is that the lunar globe as a whole must be very much more rigid than the Earth—in effect, about 1000 times more so if the ages of lunar mascons are to be of the order of 10^9 years. On the other hand, we anticipate (from known densities of lunar material) that the chemical composition of the lunar crust is not greatly different from that of the terrestrial mantle. If this is the case, the only way to endow the Moon with a rigidity sufficient to support its mascons for astronomically long times is to cool its material considerably below the temperatures encountered in the terrestrial mantle, for the most effective way to stiffen the strength of the rocks is to cool them.

It is not only the crust of the Moon that is rigid enough to support the mascons (and other, more global) departures from isostasy. The seismic evidence obtained since 1969 by three successive Apollo missions has disclosed that its material behaves like a rigid body at much greater depths, and transmits transversal 'moonquake' waves from a depth of more than 800 km. Seismometers deposited on the Moon by Apollos 11–15, and operative almost continuously for the past three years, have shown the Moon to be seismically very much more quiet than the Earth—only about

10^8 joules of seismic energy (equivalent to that of about 200 tons of TNT) gets dissipated in the Moon through seismic waves per year (in contrast with 5×10^{17} joules per year for the Earth).

Apart from a general seismic background of very low noise level, occasional moonquakes have been registered by the Apollo seismographs—due partly to external (meteoritic impact) and partly to internal causes. One of the most interesting of the latter are tectonic moonquakes emanating repeatedly from the same epicentre (of lunar coordinates $\lambda = 28°W$, $\beta = 21°S$) at a depth of 800 ± 40 km below the surface of the Moon (i.e. almost half way between the centre and the surface). A number of additional centres of moonquakes have already been detected, and some of them found to give rise to whole swarms of such quakes in rapid succession. Their activity was, moreover, found to fluctuate with varying distance from the Earth in the period of exactly one month. The frequency of seismic events appears to be maximum when the Moon is closest to the Earth—a fact which suggests that bodily tides may trigger deep-seated tectonic instabilities of a nature as yet unspecified. However, the magnitude of such events is really tiny in comparison with current terrestrial standards. The energy of individual moonquakes is, on the average, less than that liberated by the explosion of 1 kg of TNT. Virtually thousands of events so small occur daily on the Earth, and can stand out from background noise only because the Moon as a whole is seismically so quiet. The fact that such tiny moonquakes can be detected at all, and their records transmitted through the intervening gap of space, represents a veritable triumph of field seismology.

The seismic 'signature' of the observed moonquakes is very different from what we observe on the Earth: for while the terrestrial seismic disturbances are damped out in a few minutes, lunar tremors—whatever their origin—have been found to persist for 60–100 minutes. Such long durations can be understood only if the seismic waves are scattered strongly in the (mainly) basaltic crust of the Moon—about 20 km deep—which must be highly fragmented (regolith*). Between 20 and 70 km in depth the seismic velocities are indicative of the presence of solid layers of pyroxene-rich eclogite, while the mantle below the crust may consist mainly of dunite and other similar silicates. The occurrence of moonquakes at greater depths (around 800 km) implies that the lunar interior at these depths is rigid enough to support appreciable stress and transmit transversal waves—neither of which could be true if the internal temperature at these depths were to approach the melting point of rocks.

Such an interpretation is in agreement with the measurements made by the U.S. Explorer 35 satellite of magnetic interaction—or, rather, the lack of it—between the lunar globe and the solar wind. It appears that the Moon merely casts a geometrical shadow in the solar plasma, and behaves like an insulator rather than a semi-conducting body; there is an upper limit to its global electrical conductivity (indicated by a lack of any interaction) of the order of 10^{-5} S m^{-1}. This can, however, be true of basaltic material with a temperature of less than 1000°–1200°C, much lower than that prevalent in most of the terrestrial mantle. Such a temperature is sufficiently low to allow the lunar globe to have the requisite degree of rigidity but it gives very little encouragement to anyone expecting to find on the Moon suitable conditions for any large-scale volcanism. The existence of local volcanic pockets cannot, of course, be ruled out by such general arguments. Nevertheless, these arguments suggest that any volcanic activity on the Moon should have occurred on a much smaller scale than it has on the Earth.

That this should have been so could be surmised on even simpler and more general grounds. In our outline of a thermal history of the Earth in Chapter 3 we mentioned that a temperature prevailing in the interior of a planet should reflect a balance between the amount of the production of radiogenic heat in the interior of the respective body, and its loss through the surface by radiation (or other processes). The amount of heat generated in the interior should then be proportional to the volume of the respective configuration, and the loss to the area of its surface. For the spherical configuration, the ratio of the volume to its surface should be proportional to its radius. Therefore, a planet of the size of the Earth should be able to bottle up in its interior four times as much heat as the Moon, and maintain it at a considerably higher temperature than that prevailing in the interior of a smaller body—an expectation which seems to be borne out by the indirect evidence mentioned above.

The work of Explorer 35 provided us also with another piece of information characterizing the physical properties of the lunar globe: namely, the well-nigh complete absence of any *magnetic field* of our satellite. That any magnetic field of the Moon must be very weak had already been shown by experiments performed by the Russian Luna 2 in September 1959. As a result of the more recent work by Explorer 35 we know, however, that the magnetic field of the Moon, if any, does not exceed 10^{-5} A m^{-1} in strength; and that the magnetic moment of the lunar globe is probably less than one part in 10^6. This is, of course, a figure for the general magnetic field of the Moon; local fields several times stronger were found to be frozen in several samples of rocks brought to the Earth by the Apollo missions (in particular, by Apollo 12 from the Oceanus Procellarum); but such fields were probably local, rather than global. The more general result confirms—if any further confirmation were needed—that the Moon does not possess any appreciable metallic core.

The lunar surface

Let us emerge from the interior of the lunar globe to inspect its *surface* at a closer range and try to understand the diverse landmarks visible upon it. Even when seen by the naked eye from the Earth the surface of the Moon appears to be covered with bright and dark markings, which were associated by our ancestors with various popular myths. A closer look through a telescope discloses an almost bewildering array of formations whose landscape consists essentially of two principal types of ground. One, rough and broken, is comparatively bright (reflecting, in places, as much as 18 per cent of incident sunlight). The other is darker (reflecting, on the average, only 6–7 per cent of illuminating sunlight), much smoother, and superficially so flat as to simulate the surface of a liquid. The first type of ground is generally referred to as the lunar continents. They occupy large continuous areas—almost all of the far side of the Moon—and even on the side visible from the Earth they spread over more than two-thirds of its surface. The flatlands, or maria as they were misnamed by early observers of the Moon, occupy the rest. They are, on the whole, quite uniform in reflectivity and general appearance. The reader should, however, be on his guard not to associate too readily the continents with highlands or the maria with lowlands; for the detailed shape of the Moon is quite complicated, and actual elevation is only weakly correlated with different types of ground.

A more detailed telescopic inspection of the Moon discloses that the dominant formations covering both continents and maria ground alike are ring-like walled enclosures commonly called craters (see Plate 11). These occur almost everywhere on the Moon (though in greater numbers on the continents than in the maria), and are primarily responsible for giving the lunar surface its pock-marked appearance. The largest of them (mainly on the Moon's far side) attain dimensions of 300–400 km. Those with diameters in excess of 1 km number more than 300 000 on the visible hemisphere of the Moon alone, and several times as many on its far side; those smaller still are too many to be counted.

No two craters on the Moon are exactly alike. However, apart from individual distinguishing features, they have many characteristics in common; the heights of their ramparts are, in general, very small in comparison with their overall dimensions, their floors are depressed below the level of the surrounding landscape (see Plate 11), and their distribution over the surface appears to be largely random. These (and other) properties of such

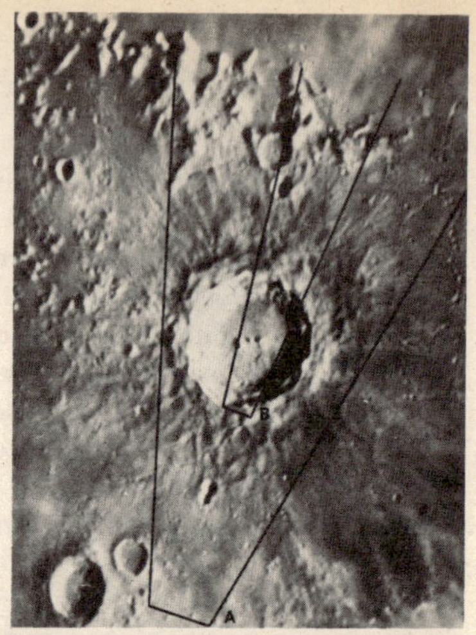

PLATE 11: The lunar crater Copernicus as photographed from the Earth (*upper right*), in comparison with views of the same formation from Lunar Orbiter 2 on 23 November 1966 from closer proximity. Photograph on the upper left shows an oblique view of this crater taken from location marked on the terrestrial photograph by A, while the one below shows a high-resolution view of its interior from vantage point B.

formations disclose that the origin of most is to be sought in impacts on the lunar surface of other cosmic bodies populating interplanetary space, the orbits of which have happened to intersect the Moon's path in the course of time.

The interplanetary space through which the Earth and the Moon continue to revolve around the Sun is not entirely empty; it contains a wide variety of ingredients of all masses and sizes, from the elementary particles evaporating from the Sun and streaming past us in space in the form of the solar wind*, through microscopic specks of dust and larger meteoritic debris to major meteorites, asteroids, or comets (see Chapters 8–10); and their orbits may occasionally intersect the path of the Moon to bring about collisions.

Of course, other planets are equally exposed to such collisions; and, in particular, the Earth should suffer an equal number of them per unit area of its surface in the same time. It is true that small particles cannot reach the Earth's surface through its protective atmospheric mantle, but our atmosphere is incapable of providing any protection against cosmic bodies whose impacts could produce crater formations kilometres in size and such formations should, therefore, accrue on the surface of the Earth and the Moon at the same rate. If the ubiquitous pockmarks on the face of the Moon have originated in this way, where are the corresponding formations on the Earth?

Such formations are indeed known to us—witness the well-known Canyon Diablo crater in Arizona, or the Tunguzka formation in the Siberian taiga. The former was caused by impact of a metallic meteorite in pre-historic times, the latter by the collision of a small comet with the Earth in 1908. The relative scarcity of other such formations elsewhere on the Earth is easily explained by a continuous operation of erosive processes by air and water, which will obliterate any surface trace of such scars within periods of the order of 10^6 years—a fleeting moment in the long geological history of our planet. The complete absence of air and water on the Moon, now or in the past, renders its surface a truly cumulative scoreboard of this kind of celestial bombardment.

However, it is impossible to explain in this way the apparent absence on the Earth of impact craters 100–300 km in size, comparable with the largest craters on the visible face of the Moon: the origin of these should have disturbed and fractured surface strata down to a depth inaccessible to the healing processes of air and water. The oldest sedimentary layers found on the surface of the Earth go back $3{\cdot}4$ to $3{\cdot}6 \times 10^9$ years in time, but show no evidence of impact fractures over areas which would be covered by them on the Moon. The only conclusion one could draw from this fact in the pre-

Apollo era was a surmise that most of the bombardment which disfigured the face of the Moon occurred before the oldest extant sedimentary beds were laid down on the Earth, and radioactive dating of the rocks brought from the Moon since 1969 has fully confirmed this surmise. As we shall see later in this chapter, the overwhelming number of such rocks are indeed more than $3 \cdot 6 \times 10^9$ years old, and solidified long before the oldest rocks known to us on the Earth.

It is therefore more than probable that the principal period of cosmic bombardment which mutilated the lunar surface extended over less than the first billion years of the existence of our satellite, and was not far removed from the time of the origin of the solar system as a whole. On the Earth and other terrestrial planets all landmarks of comparable age must have been completely obliterated aeons ago by the joint action of their atmospheres and oceans, not to speak of internal heat at deeper layers of their crusts. However, as any geological change on the Moon can proceed only at an exceedingly slow rate, its present face still bears scars and traces of many events which took place in the inner precincts of the solar system not long after its formation. If so, this should make our Moon the most important fossil of the solar system and a correct interpretation of its stony palimpsest should bring rich scientific rewards.

The near certainty that most large lunar craters, and probably also the maria (at least the mascon-possessing formations of this type) are of impact origin, and go back by age to the earliest chapter of the history of our satellite, does not preclude the existence of other craters that may be of internal (volcanic) origin. Such craters probably do exist on the Moon, but in limited numbers. There is, however, no evidence that what we mean by the terrestrial term *volcanism* ever existed on the Moon on a scale larger than on the Earth (if anything, the opposite is likely to be true) and homologues to such well-known terrestrial volcanoes as girdle the Pacific basin—or even of the European Etna or Vesuvius—are conspicuous on the Moon by their absence. We find, to be sure, other formations on the lunar surface, such as the domes or wrinkle ridges in the plains of the maria, which are indubitably of internal origin, though they are quite unfamiliar to us on the Earth; but the mechanism of their formation still remains largely unknown.

There are, on the other hand, other typical formations on the terrestrial surface which are conspicuous by their absence on the Moon. Its entire surface shows no evidence of mountain chains such as would have been formed by the folding of the crust. Such mountains as we do find on the Moon all encircle (at least partly) great mare plains, and constitute the partly destroyed ramparts of large impact craters formed in the earliest part

TABLE 7

Returning lunar spacecraft of 1968–72

Spacecraft	*Origin*	*Dates of mission*	*Nature of mission*	*Selenographic location of landing place*		*Number of hours spent on lunar surface*
				Longitude	*Latitude*	
Zond 5	U.S.S.R.	1968 Sept 15–21	unmanned (orbiter)	—	—	—
Zond 6	U.S.S.R.	1968 Nov 10–17	unmanned (orbiter)	—	—	—
Apollo 8	U.S.A.	1968 Dec 21–27	manned (orbiter)	—	—	—
Apollo 10	U.S.A.	1969 May 18–26	manned (orbiter)	—	—	—
Apollo 11	U.S.A.	1969 July 16–24	manned (lander)	23°·5E	0°·7N	2·5
Zond 7	U.S.S.R.	1969 Aug 8–14	unmanned (orbiter)	—	—	—
Apollo 12	U.S.A.	1969 Nov 14–24	manned (lander)	23°·3W	2°·5S	31·6
Luna 16	U.S.S.R.	1970 Sept 12–24	unmanned (lander)	56°·3E	0°·7S	26·4
Apollo 14	U.S.A.	1971 Jan 31–Feb 9	manned (lander)	17°·5W	3°·7N	35·3
Apollo 15	U.S.A.	1971 July 26–Aug 7	manned (lander)	3°·7E	26°·4N	67
Luna 20	U.S.S.R.	1972 Feb 14–21	unmanned (lander)	56°·5E	3°·6N	51
Apollo 16	U.S.A.	1972 Apr 16–28	manned (lander)	15°·5E	9°·0S	72

of the lunar history. Folding, the most important orogenic process on the surface of our own planet, seems to have been completely absent on the Moon, no doubt because of the greater rigidity of the lunar globe. The seismometers deposited on the Moon as a part of the Apollo 11–16 missions (see Table 7) failed to detect any indication of tectonic activity on the Moon, with instruments which, on the Earth, would have recorded thousands of such events in the same period of time. The bulk of the Moon's mass is apparently too cool and, consequently, too rigid to be susceptible of tectonic motions in the crust and the morphology of the lunar surface—based on all other aspects of observational evidence available to us so far—is entirely consistent with this view.

Our main concern in this section has so far been with the large-scale formations of the lunar surface, and what they can tell us about the cosmic environment to which the Moon has been exposed throughout its long astronomical past. The structure of this surface itself, and its chemical as well as mineralogical and petrographic composition, has proved equally revealing. Extensive studies of radar reflections from the lunar surface (conducted since 1946) have indicated that this surface down to a considerable depth cannot be an unbroken solid, but must border on space through a fractured layer of fragmented solid debris which the geologists call a regolith (see Plate 12). The origin of this regolith is no doubt to be sought in the accumulation of mechanical damage which the Moon must have suffered in the source of its long past by countless encounters with solid bodies and particles in interplanetary space.

The actual depth of such a regolith may vary from place to place; but its global average, according to the testimony of seismometers deposited on the Moon by the Apollo 11–15 missions (see Table 9) must be considerable. Since 1969, these seismometers have telemetered to the Earth the records of several moonquakes due to various causes (ground subsidence in the neighbourhood of the instrument, or impacts by spacecraft or meteorites). As already mentioned, the decay of seismic vibrations lasts an inordinately long time: in all cases more than an hour, in contrast with a few minutes on the Earth. This can scarcely be explained except by scattering of the seismic waves in a fragmented surface layer extending in depth for several kilometres.

The past history of the lunar surface makes it clear to us why this must be so; and a cosmic body of the size of the Moon or Mercury (and, to a slightly less extent, Mars) must interface with the outer space through such a regolith. This is as natural a boundary condition for a globe of that size as is the atmosphere (or hydrosphere) for a planet like Venus or Earth. The

PLATE 12: Lunar landscape from the ground—photograph taken by Neil Armstrong in the proximity of the landing place of Apollo 11 in Mare Tranquillitatis on 21 June 1969.

primary process of production of a regolith—high-velocity encounters with meteorites, asteroids, and comets—is not operative on the Moon alone; every other planetary globe is likewise exposed to it. However, as we have seen, the combined action of air and water would obliterate the scars caused by cosmic bombardment very much more rapidly than they are inflicted; so that the terrestrial lithosphere need not be (and is not) fragmented to the same extent.

The chemistry of the lunar surface

So much for the physical structure and consistency of the lunar surface layer, but what is the chemical and petrographic constitution of its material? Just as a proper understanding of its macroscopic features had to await the photographic material supplied by the lunar orbiting satellites of 1966–8 (see Table 8), a closer acquaintance with microscopic properties of lunar surface rocks would have been impossible without the contributions of the soft-landing space missions of the same period (Table 9), followed by missions, both manned and unmanned (Table 7), which returned to the Earth with actual samples of lunar material.

The atomic and molecular compositions of the material brought back from the Moon by various space missions, not very different among themselves, have been listed in Tables 10 and 11. The leading constituents are (as in the terrestrial mantle), oxygen and silicon, which in the form of silica constitute over 40 per cent of lunar surface material by mass; followed by iron oxide (around 20 per cent), aluminium oxide (11 per cent), calcium oxide (10 per cent), titanium dioxide (9 per cent), magnesium oxide (8 per cent), and other constituents amounting to 1 per cent or less by weight.

How did these results compare with previous expectations? That the surface of the Moon is covered with loosely packed material of very low thermal conductivity (and radar reflectivity) we knew before the advent of space missions, from Earth-based astronomical observations, and its relatively low static bearing strength ($3 - 6 \times 10^3$ N m^{-2}) was confirmed by the soft-landing Surveyors and Lunas (Table 9), before the advent of the Apollos. However, as regards the chemical composition of lunar material, before the last three Surveyors (which performed the first chemical analyses of the lunar ground *in situ*), and subsequent return of the Apollo samples, there were two schools of thought. One expected the Moon to possess a chemical composition akin to that of the terrestrial mantle (which should have been the case if the Moon was ever detached from the mass of the Earth); while the other, regarding the Moon an object possibly older than our planet, anticipated a similarity between the composition of the Moon

TABLE 8

Orbiting lunar spacecraft in 1966–71

Spacecraft	*Origin*	*Injection into lunar orbit*	*End of mission*	*Orbital period (minutes)*	*Number of revolutions around the Moon*	*Altitude above lunar surface at periselenium and aposelenium*	*Inclination of orbit to lunar equator*
Luna 10	U.S.S.R.	1966 Apr 3	1966 May 30	178·3		349–1017	71°·9
Orbiter 1	U.S.A.	1966 Aug 14	1966 Oct 29	208·6	547	56–1853	12°·0
Luna 11	U.S.S.R.	1966 Aug 28	1966 Oct 1	178		159–1200	27°
Luna 12	U.S.S.R.	1966 Oct 26	?	205		105–1740	4°
Orbiter 2	U.S.A.	1966 Nov 10	1967 Oct 11	208·4	2289	50–1853	11°·9
Orbiter 3	U.S.A.	1967 Feb 8	1967 Oct 9	208·6	1843	55–1847	20°·9
Orbiter 4	U.S.A.	1967 May 8	1967 Oct 6	721	225	2706–6114	85°·5
Explorer 35	U.S.A.	1967 July 19	in orbit	684	—	763–7670	11°·2
Orbiter 5	U.S.A.	1967 Aug 5	1968 Jan 31	510·3*	1201	99–6066	85°*
Luna 14	U.S.S.R.	1968 Apr 10	?	160		159– 871	42°
Luna 15	U.S.S.R.	1969 July 16	1969 July 21	150*	*126°**	132– 287	126°*
Luna 18	U.S.S.R.	1971 Sept 7	1971 Sept 11	119	54	96– 101	35°
Luna 19	U.S.S.R.	1971 Oct 3		121·8		139–	40°·6

* Changed several times in the course of mission.

TABLE 9

Soft-landing lunar spacecraft of 1966–70

Spacecraft	Origin	Date and time of landing: Date	Time (U.T.)	Selenographic coordinates of landing place: Longitude	Latitude	Gross weight (kg) (net weight on landing)
			h m s			
Luna 9	U.S.S.R.	1966 Feb 3	18 44 52	64°·37W	7°·08N	1583(100)
Surveyor 1	U.S.A.	1966 Jun 2	6 17 37	43°·21W	2°·45S	990(292)
Luna 13	U.S.S.R.	1966 Dec 24	18 1	62°·05W	18°·87N	1580(100?)
Surveyor 3	U.S.A.	1967 Apr 20	0 4 53	23°·34W	2°·94S	1040(302)
Surveyor 5	U.S.A.	1967 Sep 11	0 46 44	23°·18E	1°·41N	1006(303)
Surveyor 6	U.S.A.	1967 Nov 10	1 1 6	1°·37W	0°·46N	1008(303)
Surveyor 7	U.S.A.	1968 Jan 10	1 5 30	11°·41W	41°·01S	1010(305)
Luna 17	U.S.S.R.	1970 Nov 17	3 47	35°W	38°·28N	

and that of the solar atmosphere. The latter should approximate, as closely as anything within observational (i.e. spectroscopic) reach, to the unadulterated composition of the primordial matter from which the solar system originated: unadulterated because, on the one hand, the large mass of the Sun would have prevented any selective escape of the elements from its gravitational field, and, on the other hand, its temperature is too low to cause nuclear transformations on any appreciable scale.

TABLE 10

Mean elemental composition of the lunar and terrestrial crusts

(per cent by weight)

Element	*Earth*	*Moon*
Oxygen	46	43
Silicon	28	20
Aluminium	9·1	7
Calcium	5·0	6
Magnesium	4·0	4
Iron	3·6	15 (+)
Sodium	3·0	0·3 (—)
Potassium	1·3	0·1 (—)
Titanium	0·6	4 (++)
Carbon	0·08	<0·01 (—)
Manganese	0·05	0·03
Nitrogen	0·003	<0·01

The observational verdict delivered by the lunar spacecraft confounded both these views, and presented the Moon to us in a much more enigmatic light. For, chemically, the Moon has proved to be different from both the Sun and the Earth. For example, its content of titanium (and also chromium, zirconium, and other metals) is very much higher than in the Sun or the Earth's crust, whereas other elements (like nickel, sodium, potassium, or europium) are again very much less abundant. In particular, the ratio of iron : nickel in the Moon appears to be larger than that encountered in any other sample of cosmic matter we know (the Earth's crust, solar atmosphere, or meteorites); and the common elements like carbon or nitrogen are largely absent from the compounds found so far in the lunar crust.

What kind of rocks have been found on the Moon? All those brought

TABLE 11

Mean molecular composition of lunar surface rocks

(per cent by weight)

Molecule	*Mission*							
	Surveyor 5	*Surveyor 6*	*Surveyor 7*	*Apollo 11*	*Apollo 12*	*Luna 16*	*Apollo 14*	*Apollo 15*
Silicon dioxide	46	49	46	41	40	44	49	46
Iron oxide	12	12	4	20	21	19	20	22
Aluminium oxide	14	15	28	9	11	14	16	9
Calcium oxide	15	13	19	11	10	10	10	10
Titanium dioxide	8	4	0·4	11	4	5	2	2
Magnesium oxide	4	6	2	8	12	7	9	9
Chromium oxide				0·4	0·6	0·3	0·2	0·7
Sodium oxide	0·6	0·8	0·4	0·4	0·5	0·3	0·5	0·3
Manganese oxide				0·3	0·4	0·2	0·2	0·3
Potassium oxide				0·1	0·07	0·05	0·3	0·03

back so far are *igneous*, of generally *basaltic composition*, and numerous minerals of this type well-known from the Earth—such as olivine, plagioclase, feldspar, ilmenite, and others—have been identified in many samples. In fact, only three new minerals (not known previously on the Earth) have so far been found in lunar rocks. Their crystalline structure and chemical properties indicate that lunar rocks solidified rapidly at temperatures between 1000 and 1200°C under highly reducing conditions. The partial pressure of available free oxygen had to be less than 10^{-13} of an atmosphere to account for the virtual absence of higher states of oxidation. Moreover, many rocks brought back from the Moon exhibit evidence of shock-metamorphism, which is strongly suggestive of the effects of the passage of intense shock waves through solids, such as could be produced on the Moon by meteoritic impacts from space. Although all these lunar rocks are igneous, we should not jump to a conclusion and equate igneous with volcanic. All rocks we call volcanic on the Earth are, to be sure, igneous, but the converse is not necessarily true. Lunar rocks manifestly indicate the effects of heat treatment, but no rocks of the same structure ever passed through the crater of any terrestrial volcano!

The sources of heat which produced melting and thermal differentiation were undoubtedly local rather than global in nature. Global melting of the Moon at any time in the past is inconsistent with the shape (and gravitational field) of our satellite. A completely molten Moon would not have had time to solidify in 4600 million years; and if at least its surface has been all molten at any time, it would have solidified to a form approximating to a state of hydrodynamical equilibrium much more closely than it actually does. More localized melting could be the result of impacts, and local pockets of lava may have been produced in this way; but 'volcanoes' as we understand the term in terrestrial geology probably played a much smaller role in shaping up the face of the Moon and its composition than they did on the Earth.

The chronology of Moon rocks

The greatest and most important single piece of information furnished by mooncraft in recent years has been a determination, by chemical methods, of the *age* of the rocks from different localities on the lunar surface—or, more specifically, of the time which has elapsed since these solidified. When radiometric methods were applied to determine this age of crystalline rocks which the Apollo 11 astronauts brought back from Mare Tranquillitatis in July 1969, it proved to be close to 3700 million years. The rocks from Oceanus Procellarum brought back by Apollo 12 in November of the same year were found to be 100–200 million years younger; and Mare

Foecunditatis visited by the (unmanned) Luna 16 in September 1970 proved to be 3400 million years of age. The semi-mountainous Fra Mauro region on the fringes of Oceanus Procellarum—the target of Apollo 14 mission in February 1971—was found to be 4000 million years old; and the Apennine shores of Mare Imbrium—the landing place of Apollo 15—is 3300 million years old. Therefore, the principal pattern of lunar maria as we see them today was formed between 3300 and 4000 million years ago, as a result of an apparently unrelated series of distinct events that occurred within this time.

However, perhaps the most interesting result which has emerged from radioactive dating of lunar rocks so far is the fact that, on each landing site, the smaller-size debris (the 'fines') were found to exhibit substantially greater age than the lumps of rock. The time which has elapsed since the solidifications of the fines proved, in fact, to cluster around 4600 million years in all localities visited so far; and this age is virtually identical with that of the oldest known meteoritic material intercepted by the Earth. These fines provide an irrefutable testimony that solid matter existed on the lunar surface as far back as 4600 million years ago, and has not been re-melted since.

Nothing of comparable age has been found anywhere on the Earth, where the oldest rocks preserved in rare localities are not much more than 3600 million years old (all older strata having been consumed by the interior). In other words, the first 1000 million years of the Earth's history represent the 'dark aeon', of which we find no testimony engraved in the terrestrial stony strata. Since 1969 we have, however, unlocked a source of material which has suddenly illuminated for us the earliest chapter in the history of the Earth–Moon system, and is enabling us to reconstruct an almost uninterrupted story of what has happened in the inner precincts of our solar system since the days of its formation.

In contrast with the Moon, our mother planet is exhibiting to the outer space a cosmic face of almost eternal youth—rejuvenated continuously by modification of its surface by the joint action of air and water, or (more important) by continuous drifts in its mantle driven by the heat engine in the interior. Indeed, a very small fraction of the terrestrial continents or ocean floors is more than a few hundred million years old; and this is less than ten per cent of the age of our planet.

Unlike the Earth, the Moon—on account of its small mass and heat capacity—can afford none of these cosmetic agents. As a result, the Moon preserves a cumulative record of events that took place long before our terrestrial continents were formed, or before the first manifestations of life flickered in shallow waters on the surface of the Earth.

8
The Asteroids

THE MAJOR AND terrestrial planets, the principal properties of which have been described in the preceding two chapters, are not the only celestial inhabitants of our solar system. Indeed most of these have yet to be introduced in our narrative; and although they do not add up to much in terms of mass, the number of bodies still to be accounted for in the total census is enormous. We can divide this remaining population into two distinct groups: asteroids, which are covered in this chapter, and comets which are discussed in Chapter 9.

The asteroids (or planetoids, as they should be called, for they have nothing to do with the stars) occupy a significant position within the framework of our solar system by their numbers as well as by the space which they traverse. Johannes Kepler—discoverer of the true model of the planetary system in the early part of the seventeenth century—was struck by an apparently empty wide gap of space between Mars and Jupiter. He surmised that a new planet might one day be discovered which would fill this gap, and in a way this has proved to be the case.

The first indication that the space between Mars and Jupiter was not entirely empty was not found until the nineteenth century. On 1 January 1801, the Italian astronomer Giuseppe Piazzi of Palermo noted in the constellation of Taurus a small star of eighth apparent magnitude, which was changing its position fairly rapidly among others. Mindful of the discovery of Uranus in 1781, which was then recent history, he suspected at once that he had found a new planet.

This discovery had a rather dramatic sequel. Illness soon prevented Piazzi from continuing his observations, and by the time he recovered sufficiently to return to his telescope, the region of the sky in which his object would be found had disappeared in evening twilight and could not be observed again for several months—until it re-emerged from behind the Sun in the morning sky. Where, however, should he look after so long a time lapse? The problem thus posed gave a chance for a brilliant young mathematician, Carl Friedrich Gauss, to show his mettle: by an application of the method of least squares (discovered by him a few years before) he succeeded in extracting from Piazzi's observations sufficient orbital elements of the new body to predict its future position. His ephemeris* enabled J. G. Olbers to re-discover it on 1 January 1802—exactly one year after it was first spotted by Piazzi.

A comical element was introduced into this drama by G. W. F. Hegel, a budding German philosopher (then 31 years of age), who on 27 August 1801 defended a thesis (*Dissertatio philosophica de orbitis planetarum*) to qualify for a university lectureship, proving (according to his lights) that there could not be more than seven planets in the sky—eight months after Piazzi had found the eighth! News travelled more slowly in those days than it does today, but Piazzi's discovery was known to German astronomers in March of 1801, and Hegel's ignorance earned his thesis well-deserved contempt* as the last disastrous attempt to regulate the number of celestial bodies by naïve philosophical preconceptions.

Hegelian nonsense notwithstanding, the celestial object discovered by Piazzi in 1801 proved to revolve around the Sun in a mildly eccentric orbit ($e = 0{\cdot}08$) close to the plane of the ecliptic ($i = 10°{\cdot}6$) in a period of 4·6 years, almost exactly half way between Mars and Jupiter. It was given the name of Ceres. It did not remain the only known celestial body of its class for long. In March 1802 Olbers discovered another such object—called by him Pallas—which proved to revolve between Mars and Jupiter in almost the same period as Ceres. It travelled in a much more eccentric orbit ($e = 0{\cdot}24$), inclined by 35° to the plane of the ecliptic, but approached Ceres closely at one point. Olbers conjectured that Ceres and Pallas were originally parts of the same parent celestial body (an assumption held by some even now). Watching for possible additional splinters that might pass near the common point of intersection, Harding discovered in 1804 a new such asteroid (Juno), and in 1807 Olbers detected another (Vesta) which did not, however, approach the other three.

After 1807, almost forty years elapsed until a new celestial object of this class (Astraea) was discovered in 1845, but since then the series of new

* The Duke Ernest of Saxony Gotha sent a copy of Hegel's thesis to the astronomer Franz Xaver von Zach with the inscription *Monumentum insaniae saeculi decimi noni*. But the great mathematician Gauss wrote later (in a letter to a friend) that Hegel's dissertation of 1801 *de orbitis planetarum* was still a jewel of wisdom in comparison to some of his later philosophy.

discoveries has been almost uninterrupted—especially since the advent of astrophotography in the latter part of the nineteenth century. By 1900, over 500 asteroids were known and since that time (especially since Schmidt photographic telescopes came to bc used) the number of known objects of this class has grown by leaps and bounds. Asteroids with known orbits now add up to more than 1800, while the number of those within reach of our photographic telescopes (i.e. attaining 20th apparent magnitude at the time of their opposition) probably exceeds 50 000; those still smaller are beyond count. The physical and kinematic characteristics of some of the largest asteroids are given in Table 12.

By the distribution of the semi-major axes of their orbits (Figure 5) the asteroids are spread almost through the entire space-gap between Mars and Jupiter (i.e. between 2 and 5 AU). Some of them (the Trojans, a group bearing the names of the Homeric heroes of the Trojan war) closely trail Jupiter's motion in the same period and at the same distance, while others almost skirt the orbit of Mars. The distribution of their semi-axes between 2 and 5 AU is highly uneven (Figure 5) and exhibits a conspicuous clustering into groups of objects with orbits similar not only in size, but also in shape and orientation. Ceres and Pallas constitute the core of one such group, and so do the Trojans; asteroids of the Hilda or Themis groups are even more numerous.

Just as certain favoured distances from the Sun predominate among asteroids, others are apparently unpopular, thus forming gaps* in the frequency-distribution of the semi-major axes of asteroidal orbits, corresponding to certain commensurabilities between the mean motion of the asteroid and Jupiter (the ratios of 1 : 3, 2 : 5, 3 : 7, and 1 : 2 being particularly conspicuous). The reason for these particular zones of avoidance is dynamical, and is explained by the vulnerability of orbits in such gaps to planetary perturbation; a mass-particle that happens to drift into such zones will relatively soon have its orbit transformed by perturbations so as to fit into a more stable group outside the gaps.

The eccentricities of asteroidal orbits range widely, and some (for instance, Icarus, for which $e = 0{\cdot}83$, or Adonis with $e = 0{\cdot}78$) are of cometary rather than planetary orders of magnitude. Such asteroids traverse a wide range of interplanetary space in their revolution around the Sun. Thus while Icarus (with an orbital semi-major axis A of only 1·08 AU) can approach the Sun within less than half the distance of Mercury at its

* These are usually known as Kirkwood's gaps, although their existence was first pointed out by the Prague astronomer K. Hornstein, whose work antedated Kirkwood's but was generally overlooked.

TABLE 12

Kinematic and physical properties of the principal asteroids

Number (in order of discovery)	Name	Year of discovery	Orbital period (years)	Semi-major axis (AU)	Orbital eccentricity	Orbital inclination	Apparent visual magnitude in opposition	Mean diameter (km)	Period of rotation (hours)
1	Ceres	1801	4·60	2·77	0·080	10°·6	$7^{m}.4$	770	9·1
2	Pallas	1802	4·61	2·77	0·239	34°·8	8·0	490	10·1
3	Juno	1804	4·36	2·67	0·257	13°·0	8·7	200	7·2
4	Vesta	1807	3·63	2·36	0·089	7°·1	6·5	410	10·6
5	Astraea	1845	4·13	2·58	0·186		9·9	135	16·8
6	Hebe	1847	3·77	2·42	0·202	14°·8	7·0	170	7·3
7	Iris	1847	3·69	2·39	0·231	5°·5	6·7	170	7·1
8	Flora	1847	3·27	2·20	0·157		8·8	140	13·6
9	Metis	1848	3·69	2·39	0·123		8·1	190	5·1
10	Hygeia	1849	5·60	3·15	0·100	3°·8	8·5	260	18·0
15	Eunomia	1851	4·30	2·64	0·187	11°·8	7·4	270	6·1
16	Psyche	1852	4·99	2·92	0·139	3°·1	8·6	220	4·3
20	Massalia	1852	3·74	2·41	0·143		8·2	160	8·1

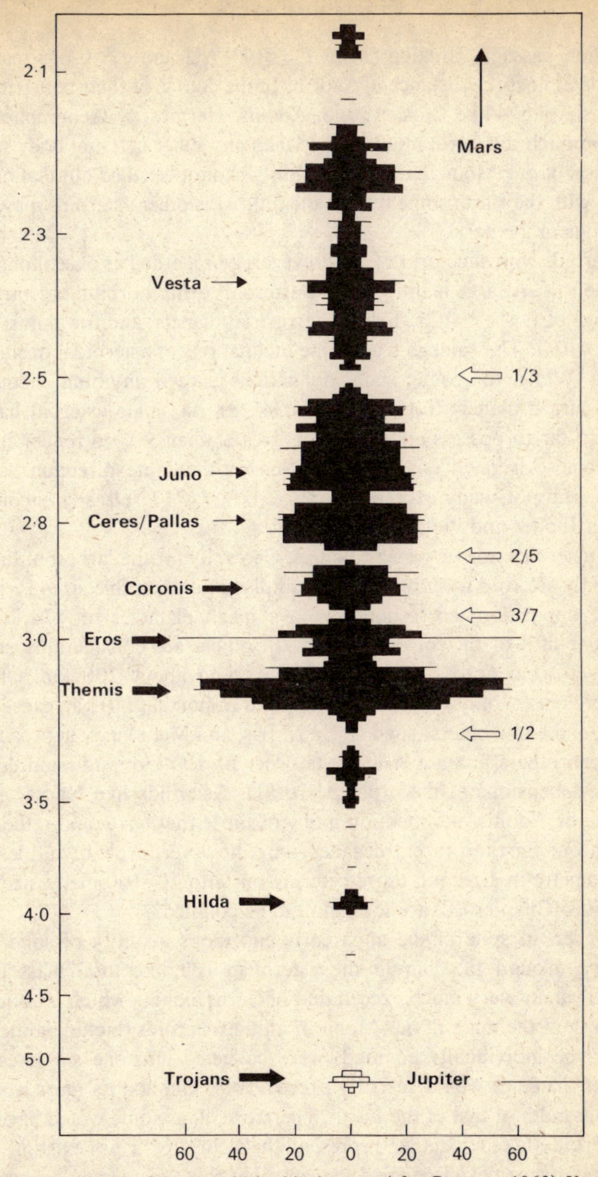

FIG. 5: Size distribution of the asteroidal orbits in space (after Brouwer, 1963). Vertical axis: orbit size (AU); horizontal axis: number of orbits of the respective size.

perihelion passage, Hidalgo (with $A = 5{\cdot}71$ AU and $e = 0{\cdot}65$) reaches at aphelion almost the distance of Saturn. In the course of their peregrinations, some asteroids—like Eros, Apollo, Adonis, Hermes, or Geographos—can also approach the Earth more closely than any other celestial body with the exception of our Moon, and the possibility cannot be ruled out that one may collide with the Earth some time in the future (as other asteroids may indeed have done in the past).

Asteroids with such orbits can, however, be regarded as exceptions rather than the rule. A large majority of the asteroids exhibit orbital eccentricity in the range of $0 < e < 0{\cdot}25$, like the terrestrial planets; and for almost half of them $e < 0{\cdot}1$. The same is true of the inclinations of asteroidal orbits to the ecliptic. While, for some, such inclinations exceed anything encountered among larger planets (for Pallas, $i = 34^\circ{\cdot}8$), no single asteroid has been found so far to possess an inclination greater than 90° to render its orbit retrograde. All revolve around the Sun in the same direction as other planets, and not many exceed the range $1^\circ < i < 17^\circ$ already encountered between Jupiter and Pluto.

The asteroids are very small bodies. Ceres, by far the largest of them all, is the only asteroid to show an apparent disc of measurable size ($1''{\cdot}1$ at the time of opposition) corresponding to a mean diameter of 770 km—i.e. about one-fifth of that of our Moon. All others are considerably smaller; Pallas, 490 km across; Vesta, 410 km; Juno about 200 km. Of those discovered later, only Hygeia (260 km) and Eunomia (270 km) exceed Juno in size; and a few others (such as Hebe, Iris, and Metis) may approach it. A few dozen others attain a size of the order of 100 km; and hundreds may possess dimensions of the order of 10 km. Asteroids like Icarus, Apollo, Hermes, or Adonis—to mention a few of those that have paid rather close calls on our Earth in recent decades—are, however, probably all less than one kilometre in size, and those petering out into 10–100-metre size of the meteorites (Chapter 10) are too many to be counted.

However, in spite of the apparently enormous quantity of solid debris revolving around the Sun in the asteroidal belt, the total mass of this material cannot very much exceed that of Ceres alone—which is about one-hundredth of the mass of our Moon. If all the asteroids (including those too small to be individually counted) were collected into one single celestial body, its diameter would scarcely exceed 1000 km, and its mass would be one-thousandth of that of the Earth. Therefore, if, as some would have it, all the asteroids taken together represent the debris of a pre-existing planet destroyed by some unspecified catastrophe, this planet would have been very much smaller and less massive than the Moon, let alone the Earth.

The composition of the material constituting the asteroids cannot be investigated at a distance, and can be deduced only from the composition of solid debris intercepted by our Earth in the form of meteorites, some of which may be migrants from the asteroidal belt. An analysis of such material reveals indeed a fascinating story, which will be discussed in Chapter 10.

Another interesting feature of the asteroids is their axial rotation. Cosmic (i.e. self-gravitating) bodies of asteroidal masses need not necessarily be spherical. As we mentioned already in connection with our Moon, each self-gravitating body, whether fluid or solid, is compelled by Nature to become spherical whenever hydrostatic pressure in the interior is sufficient to overcome the strength of its material. Hydrostatic pressure depends on the total amount of mass. Therefore, with increasing mass, a point is bound to be reached when any kind of material will be crushed by the weight of overlying matter.

According to Wildt, self-gravity compels every solid body of density ρ (in$\oplus$) to become spherical if its radius exceeds some $250\rho^{-\frac{1}{2}}$ km. For silicate rocks of mean density 3300 kg m^{-3} (i.e. $\rho = 3{\cdot}3/5{\cdot}5 = 0{\cdot}6$), this critical radius should be close to 320 km—much less than that of our Moon (1738 km). So large a disparity should be sufficient to enforce spherical form on our satellite almost regardless of its composition. The asteroid Ceres with a radius of 390 km appears to be still above the limit, but Pallas and Vesta are well below it and so are all the other asteroids. Therefore, if such bodies ever acquired aspheric form, they could retain it indefinitely and their present characteristics could be directly related to those at their origin.

With the exception of Ceres (and, possibly, Pallas and Vesta) the apparent discs of all asteroids are too small to enable us to discern their actual form. However, their light is generally found to be variable, and to oscillate with clock-like regularity in periods of several hours (all less than a day). For bodies so small the cause of variability can be ascribed only to variable cross-sections exposed to the external observer at different times and the amplitudes of the observed light changes indicate the existence of very appreciable departures from spherical form.

One additional interesting fact emerges in this connection: namely, that the light changes of all asteroids appear to be simple—there is no indication of the beat phenomena that would arise if an irregular-shaped body were tumbling in space like a spinning top. The fact that, judging from their light changes, asteroids appear to be rotating about a single axis whose direction is fixed in space, appears to be at variance with a view that these bodies originated in a collision of two planets (or other kind of similar catastrophic event), for such a process would have been bound to endow the resulting splinters with an arbitrary distribution of rotational momenta in three dimensions, not favouring any one particular direction.

Would it be possible for a planetary body to be destroyed in a way that would ensure that the resulting splinters would rotate about only one axis? Such a situation could have occurred if, for instance, the disruptive force were the tidal action of the Sun—or, in other words, if a new-born planet happened to wander so close to the Sun that it was torn to pieces by its attraction. In Chapter 2 we mentioned such a mechanism in connection with the possible origin of Saturn's rings. Is it possible that the particles now constituting the asteroidal belt are depleted remnants of a whole ring which once encircled the Sun, and that the Kirkwood gaps represent phenomena analogous to the Cassini division and other details observed in Saturn's rings?

This is indeed possible, but perhaps not very likely; for in order to bring the present asteroidal belt so close to the Sun, the radius of the Sun at the time would have had to exceed one astronomical unit, and the solar surface to extend beyond the present orbit of the Earth. This may once have been true, but could any planet have wandered near it from outside? Besides, the asteroidal belt (at least in its present form) is not as flat as Saturn's rings. Far from it, for the r.m.s. deviation of the inclinations of present asteroidal orbits to the ecliptic amounts almost to 10°, and such a dispersion could not have been brought about by subsequent planetary perturbations.

On the other hand, it is possible that in reality the situation was reversed, and that solar perturbations prevented the formation of a planet between Mars and Jupiter from pre-existing material. In other words, it is possible that the present particulate contents of the asteroidal belt represents a surviving sample of the legendary planetesimals from which the planets in general are supposed to have come into being. However, we do not know, and much work remains to be done before such a suggestion can be placed on a more solid footing.

9
Comets of the Solar System

OUR ACCOUNT OF the principal members of the solar system has so far omitted one enigmatic class of celestial bodies which do not add up to much as far as mass is concerned, but whose many other characteristics set them apart from the rest of the planetary population: namely, the *comets.*

We must first consider the motions of the comets around the Sun. All constituents of the solar system described in the preceding chapters conform to certain rules which circumscribe their motions within fairly narrow limits—they revolve around the Sun in the same direction, in orbits which are approximately circular, and which are situated very nearly in the same plane. The comets are the only bodies in the solar system to defy all these rules in a most flagrant manner. Their orbits are, in general, highly eccentric, and their planes are inclined virtually at random to the invariable plane* of the solar system, in which the planets revolve.

For example, while comets like Schwassman–Wachmann 1927 (named after its discoverers and the year of discovery) or Comet Oterma 1942 revolve around the Sun, with periods close to 16 years and 8 years, respectively, in orbits which are almost circular, others (like comet Delavan 1914) almost skirt the parabolic limit. Or again, while many comets revolve around the Sun in the same direction as the planets (i.e. their inclinations i to the invariable plane, while possibly large, are less than 90°), others are known to revolve with inclinations between 90° and 180°, so that their orbits are, in fact, retrograde. The famous Halley comet, whose periodic visitations every three-quarters of a century have become a permanent feature of the history of astronomy in the past two thousand years, belongs to this class.

Motion along such orbits can take the comets over a very wide range of distances from the Sun: from far beyond the orbit of Pluto to within the orbit of Mercury. At great distances from the Sun (generally beyond the orbit of Jupiter) a comet—if it could be observed by its proper motion on the stellar background—would possess a star-like appearance, revealing in no way its true nature. At solar distances between those of Mars and Jupiter, however, the image of a comet begins to grow fuzzy and to simulate a small diffuse nebula, tending to become elongated in a direction away from the Sun—an elongation which will eventually develop into a cometary tail.

Very few comets become visible to the naked eye, let alone develop a tail, at a distance greater than one astronomical unit away from the Sun. But the nearer they approach our central star, the brighter comets become, and the more conspicuous the tail they can spread out in the sky. Proverbial comets with bright and long tails have become scarce in this century, and probably very few readers will have seen one. The nineteenth century, however, witnessed several comets (like those of 1811, 1848, and 1858, or 1861 and 1882) which stretched their tails over a large part of the sky; and Cheseaux's comet of 1744—that peacock of the cometary tribe—spread out not less than six tails at the same time to the admiration of countless onlookers.

The great changes in brightness exhibited by comets at varying distances from the Sun strongly suggest that the origin of their light is to be sought in different manifestations of sunlight. At large distances from the Sun, in the cold storage of the outer parts of the solar system, where the temperature is below 100 K, a comet is reduced to a solid *core* which merely reflects sunlight like an asteroid. As its distance from the Sun decreases this core begins slowly to sublime and surround itself with a gaseous *coma* which starts emitting light of its own (excited by sunlight), and is dragged along by the core through space.

As the comet moves closer to the Sun, the gas liberated by evaporation can no longer be contained in the coma, and begins to escape in a direction generally away from the Sun. This escaping jet of gas, generally known as the *tail* of the comet, represents, in effect, a net *loss* of cometary material which the feeble attraction of a cometary nucleus is virtually powerless to prevent. The reason why cometary tails, sometimes extending over millions of kilometres in space, are neither straight nor simple is that competing forces are acting upon them. The force that directs them generally away from the Sun is, of course, the radiation pressure of sunlight. This is, however, not much stronger than the centrifugal force of orbital motion (which also varies with the distance from the Sun). This tends to displace cometary material in a direction at right angles to that of sunlight. Or, again,

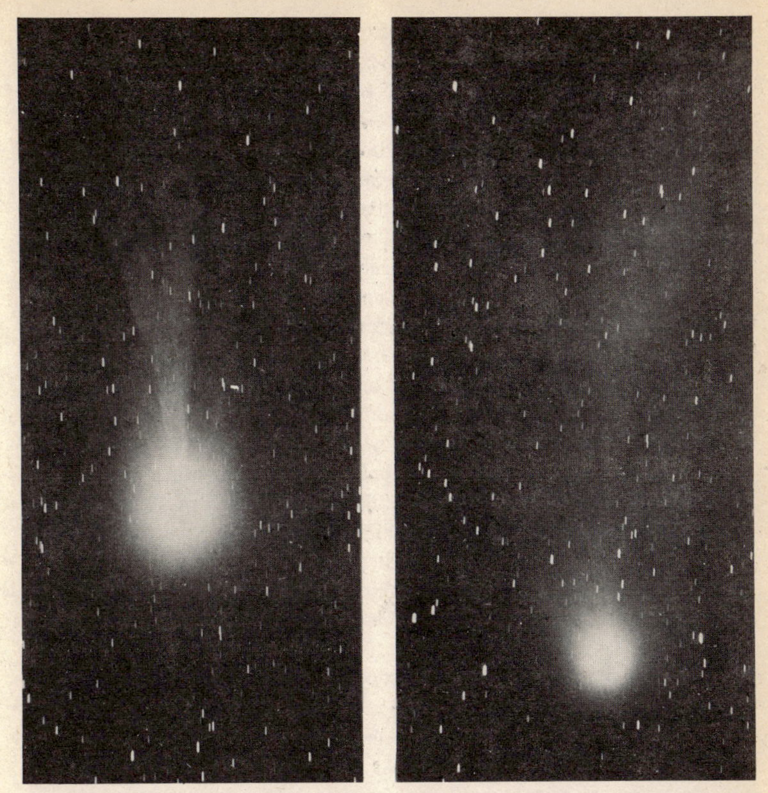

PLATE 13: Typical view of a comet—1942f (Whipple)—photographed in 27 January 1943 (*left*) and 28 (*right*) by G. van Biesbroek with the 24-inch reflector of the Yerkes Observatory. Star trails denote the proper motion of the comet in the course of each exposure.

the electrically charged atoms and molecules photoionized by sunlight may be at the mercy of interplanetary magnetic fields carried by the solar wind. The forces acting on the material of cometary tails are therefore complicated, and the great variety of tails observed in the sky bears witness to this fact.

A simplified anatomy of a comet (Plate 13) divides it into a core, its coma, and the tail. In spite of external appearances the inconspicuous cometary core is the only part of its anatomy that really matters. Its dimensions are insignificant. We know that it is barely a few kilometres across, for whenever a comet transits across the Sun it vanishes completely out of sight down to its core, with no trace of a black dot seen against the disc of the Sun. Its mass must likewise be minute by astronomical standards, for even the closest approach of a comet fails to disturb in the least the appointed motion of any other celestial body in the solar system. For example, comet Brooks passed in 1889 between Jupiter and its tiny fifth satellite (Amalthea) and was punished by being split in four parts, but Amalthea did not acknowledge the presence of the interloper by the least flicker of its own motion. From this (and other more indirect evidence) we surmise that the masses of cometary heads do not, on the average, exceed some 10^{14} or 10^{15} kg, although an exceptional comet may, perhaps, be more massive. A mass of this order of magnitude may seem large in comparison with our common terrestrial standards, but it represents only one part in 10^9 of the mass of the Earth as a whole, or one ten-millionth of that of the Moon.

Since the sizes of the cometary cores do not exceed a few kilometres, their mean densities will not exceed about 1000 kg m^{-3}. So low a mean density of the core indicates that the bulk of its mass is formed not by solid rocks but by frozen ices of more volatile substances, which with increasing proximity to the Sun begin to sublime into interplanetary space. The cometary comas and tails represent by-products of such a process. The volume of cometary tails is certainly enormous, but the mean density of gases escaping through them is only about 10–100 times higher than that of the surrounding interplanetary substrate.

What does cometary matter consist of? As long as it remains confined to the solid state of the core we have no ready way of probing its composition at a distance. However, as soon as its material evaporates to form the coma and the tail, the gas becomes luminous by virtue of solar photo-excitation, and its principal constituents can be determined by spectral analysis. Needless to say, the degree of photo-excitation varies with the distance from the Sun. The only molecular bands apparent in the spectra of the comets outside the orbit of Mars are those due to the neutral molecules of CH, CN, or C_2. Inside the Martian orbit, ionized diatomic molecules (CO^+, N_2^+, or OH^+) begin to make their appearance, together with triatomic molecules like CH_2 and NH_2. At distances smaller than one astronomical unit from the Sun, even metallic lines (sodium, iron, nickel) begin to show up in the spectrum as the cometary tails develop in their full bloom. All these elements, and more, must be present in the cometary cores, from which they are released by melting and evaporation. However, what kind of molecules—possibly much more complex—they were part of in the solid state we do not know; all that the spectroscopic analysis can disclose is the composition of gases dissociated by sunlight.

Therefore, as the comet begins to approach the Sun through the inner precincts of the solar system and starts developing a tail, it is bound to lose mass irretrievably to the interplanetary substrate. How long can a comet afford to sustain such a loss and avoid complete disintegration? The answer is hardly surprising: cosmically speaking a very short time. A more exact verdict depends essentially on how closely a comet actually approaches the Sun. Its disintegration obviously proceeds the faster, the longer it stays near the Sun. But an assessment of the amount of mass lost in the course of each orbital cycle discloses that the average lifetime of a comet that approaches the Sun within less than the distance of the Earth is limited to centuries rather than millennia.

This time represents so small a fraction of the total age of our solar system that we must ask ourselves with some concern where the comets come from, and from where are they replenished, if we are to see any around us at all at the present time? The bulk of the cometary matter is highly volatile, and chemically akin to the composition of interstellar matter or the major planets rather than to the planetary bodies of more moderate mass. Moreover, if dozens are being melted down each century by the scorching heat of the Sun, where do they come from? In what kind of a pool could they have survived $4{\cdot}6 \times 10^9$ years and avoided becoming extinct in an earlier age of the history of the solar system? It is obvious that such a pool could only be located beyond the periphery of our solar system, far away from the Sun, where the prevalent climate approaches that of interstellar space. As we already mentioned, outside the orbit of Pluto the temperature of solid material heated only by sunlight drops well below 100 K, and cometary ices kept in such cold storage could afford an almost endless lifetime.

Whether the comets constitute the left-overs from the days of the formation of the solar system, or whether they were acquired subsequently by the Sun from adjacent regions of interstellar space, we do not yet know; and both possibilities must still be kept in mind. But what force can lure them out, now and then, from the relative safety of their cold storage on to the treacherous stage of the inner precincts of the solar system? The most probable answer is the attraction of the major planets—mainly of Jupiter and Saturn. It is these planetary giants which, if the configuration is favourable, manage to persuade comets approaching them from outer space to take a sightseeing trip deeper into the solar system, and for comets which succumb to this temptation such a course is tantamount to a death-warrant, effective after they have gone around the Sun a few dozen times.

Once their orbits have been reduced to ellipses which approach the Sun within a short distance, only one event can rescue comets from impending doom: that is if, on their outward journey away from the Sun, they manage to approach a major planet whose gravitational pull enables them to put on speed and escape back into the relative safety of the cold storage in the outer parts of the solar system. Each century we see more than one comet disappear for ever from the list of our regular cosmic visitors, just as we have seen several of them disintegrate virtually in front of our eyes, and others become pale ghosts of their former glamorous selves by gradual dissipation of their mass into space.

Some 20 years ago the Dutch astronomer Oort proposed an interesting hypothesis, according to which the cometary cores are, in fact, asteroids ejected from their near-circular orbits between Mars and Jupiter by the gravitational action of Jupiter, in much the same way as Jupiter is known to have been ejecting comets up to the present time. Once an asteroid gets ejected along a highly eccentric orbit beyond the distance of Pluto, its surface may in time become encrusted with frozen ices of hydrocarbons like a snowball. It will be dispossessed of these again should such a snowball—now a comet—be lured once more, by gravitational perturbations, into the inner precincts of the solar system.

Indications that cometary nuclei may represent more than loose agglomerations of hydrocarbon ices, lacking any solid cores, have been provided by certain comets in the past (such as the great comet of 1843) which at the time of their perihelion passage approached the Sun so closely that a loose conglomeration of frozen hydrocarbons would almost certainly have been destroyed completely by solar heat. We cannot say that nothing happened to such a comet as a result of this experience, but it is certain that it survived; and from this fact the Dutch astronomer Minnaert concluded that this comet must have possessed a solid core at least half a kilometre in size.

But if we return to the balance of the rate of acquisition and loss of comets from our neighbourhood, we are led to conclude that the total supply of cometary cores (of whatever structure) still available in the cold storage which our Sun drags along with it through space must be of the order of 10^{10} to 10^{11}. Since, however, each comet weighs on the average only 10^{14}–10^{15} kg, the total mass of cometary material available in the solar system is likely to amount to no more than that of our Earth. This total mass will become larger if Oort is right in his view that asteroid-like bodies are necessary for the 'seeding' of comets and constitute, in fact, their inner cores.

10
Meteors and Dust in Interplanetary Space

IN GIVING AN account of the physical structure and chemical composition of the heads and tails of the comets we have omitted so far one essential ingredient: namely, the solid particles, stony as well as metallic, which are mixed up with cometary ices and are probably accrued by them on long cometary sweeps through space. In effect, cometary ices are decidedly dirty. As their snowballs of frozen material continue to melt and disintegrate in the scorching heat of the Sun at the time of cometary perihelia, the solid particles embedded in them will begin to fall apart again. The debris so discarded will trail in the wake of the comet until, eventually, it will clutter up most of its path. When such straggling particles are intercepted in their journey by our Earth (or, indeed, any planet), they manifest themselves to us as meteors during the last seconds of their cosmic existence as compact particles, and flare up briefly in their own light as atmospheric resistance heats them up to vaporization. What can their numbers, motions, and composition tell us about the particulate contents of interplanetary space?

Meteors of the solar system

As we have learned in earlier chapters, the most important single characteristic of any celestial body is its mass, and this applies equally to the meteors. The mass of most of the meteors which you may have seen flashing across the sky at night is very small—generally less than 1 g. The reason why a mass so small can give rise to luminous phenomena that can be seen at distances of many kilometres is because of the high velocities with which meteors may intersect the orbit of the Earth and enter our atmosphere. A particle weighing 1 g but moving with a velocity v of 30 km s^{-1} possesses a kinetic energy $\frac{1}{2}mv^2 = 4{\cdot}5 \times 10^5$ N. The atmospheric resistance decelerating such a particle does work approximately equivalent to the braking of a 700 kg car from a speed of 90 miles per hour to a standstill in a few seconds—no wonder that the brakes get completely burned out in the process!

Most meteors do not, in fact, last more than a second or two, and spend themselves at altitudes of 80–120 km in fleeting flashes of light—due partly to the incandescent material of the cosmic intruder, and partly to the glow from the atmospheric cushion compressed in front of the moving meteorite. The depth of penetration of a meteorite increases, of course, with its mass. The descent of meteorites weighing kilograms through denser layers of the lower atmosphere may be accompanied by dazzling displays of light, lasting several seconds, which can turn night into day. Some of them may actually hit the ground in the solid state to enrich the collections of meteorites in our museums.

A few times in a century the Earth encounters meteorites which penetrate the atmosphere with a residual mass of hundreds, or even thousands, of kilograms. These impact on the terrestrial surface amidst spectacular effects which attract world-wide attention. The last such fall occurred in the Sichote Alin mountains of eastern Siberia in February 1947, when a whole swarm of metallic meteorites, whose total mass exceeded 100 tons, collided with the Earth. Against such bodies our atmosphere, which consumes most of the small particles, ceases to provide any effective protection. Meteorites of moderate size may still be decelerated by it enough to survive on impact in the solid state. Bodies substantially larger, weighing thousands of tons, pay little or no attention to our air, and impinge on the ground with their original relative cosmic velocity—which may (for parabolic encounters) approach 72 km s^{-1}. In such cases, a sudden conversion, on impact, of a large part of the kinetic energy into heat may be more than enough to vaporize the entire mass of the intruder, which explodes like a bubble of hot gas. Such an explosion may leave behind it a scar on the solid surface in the form of an impact crater, though the crater itself may be largely devoid of any trace of the actual material of the cosmic intruder which was blown away in the explosion.

We have already discussed phenomena of this nature in Chapter 7 in connection with the history of the lunar surface. On the Earth, erosive processes due to combined action of air and water limit the time of survival of such astroblemes*, due to impacts of large meteorites, small asteroids, or comets, to generally less than a million years. Several dozens of these have, however, been identified by geologists in different parts of the world as proof that the days of the bombardment of planetary surfaces by cosmic projectiles of a calibre heavy enough to produce impact craters thousands of metres in size are not yet completely over.

The best-known examples of such types of formations are the meteor crater near Canyon Diablo in Arizona, U.S.A., caused by the impact, 10–50 thousand years ago, of a metallic meteorite with a mass of the order of 10^5 tons, or the Siberian Tunguzka crater of 1908 which was probably due to a collision of the Earth with a small comet. Very much larger fossil craters, probably of impact origin, have been found in geological strata deposited in more ancient times. One of these is the Vredefort dome in South Africa, which is some 100 km across and dates from Permian times.

Very little of the material of the impinging meteoritic bodies is found near such formations, because most of it evaporated at the time of the fall. However, when meteoritic material does survive in the solid state and finds its way to our laboratories, its analysis can furnish a wealth of information concerning the composition and the past cosmic history of such samples.

More than 90 per cent of all meteoritic matter recovered has proved to be from stony meteorites of essentially silicate composition with different admixture of calcium or oxidized iron. The class of stony-iron meteorites forms a natural transition to the genuine iron meteorites, the bulk of which consists essentially of an iron-nickel alloy, with a little admixture of cobalt. Needless to say, metallic meteorites of this latter class are very much more resistant to exposure on the terrestrial surface as well as much easier to identify as such than stony meteorites; as a result, the known finds probably exaggerate the actual cosmic abundance of iron meteorites over stone meteorites to a considerable extent.

Their mineralogical and petrographic structure renders the meteorites quite unlike any other samples of cosmic solid matter which we have encountered in the crust of the Earth, or on the Moon so far. In particular, the meteoritic irons show evidence of peculiar crystalline structure (Widmannstätten figures) which suggests an extremely slow rate of cooling at the time of crystallization—a rate which almost requires that the cooling took place in the interior of a cosmic body of sizeable dimensions, possibly (but not necessarily) associated with the asteroidal belt (cf. Chapter 8).

Of greater interest are the ages of the meteorites measured from the time of their solidification, as determined by the radioactive methods discussed in Chapters 3 and 7 in connection with the Earth and the Moon. For a very large majority of meteoritic samples, especially for iron meteorites, these ages have been found to cluster between $4{\cdot}5$ and $4{\cdot}6 \times 10^9$ years: close to the age established for the oldest rocks on the lunar surface (see p. 92). So close a coincidence strengthens our conviction that this age represents a very important date in the history of the solar system, probably the time of its origin (see Chapter 11).

The time that has elapsed since the solidification of meteoritic material should be clearly distinguished from the exposure ages indicated by the accumulation of nuclides originating by interaction of meteoritic material with cosmic rays (solar or galactic). Such exposure ages of the meteorites have been found to range from millions to a few hundred million years—much less than the time elapsed since the time of solidification. The difference can be explained if, before the time indicated by the exposure age, the meteorite was shielded from cosmic rays by cometary material in which it may have been embedded. For stony meteorites, one could explain it as a result of gradual disintegration of the crust by space erosion, or again through fragmentation by mutual collisions.

The parent region of meteoritic material in the solar system is usually identified with the asteroidal belt between Mars and Jupiter, 3–5 AU away from the Sun. If this is so, however, a gradual spiralling down of a particle from the asteroidal belt to the distance of the Earth by the cumulative effect of planetary perturbations alone would take several hundred million years—a time long in comparison with the observed exposure ages of most meteorites. In particular, those with exposure ages of the order of 10^6–10^7 years could have traversed this distance only if they hitched a ride on the nuclei of comets, having been swept up by them during their passage through the asteroidal belt, and tossed away again nearer to the Sun.

It is probable that a large majority of the meteorites of asteroidal origin which are being picked up by the Earth have been imported into the inner precincts of the solar system by the comets. Those which become loose by the gradual disintegration of cometary heads are bound to follow in the wake of their former carriers at least for some time, in the form of *meteoritic swarms* which loosen up gradually by depletion and planetary peturbations. Swarms which are relatively young are apt to be more compact, and the meteor showers observed at the time of an intersection (or close approach) of their trajectories with the terrestrial orbit are likely to be intensive, but of short duration. Sometimes meteoritic material revolving in them is concentrated in knots; an encounter of the Earth with one of these gives rise to short-lasting magnificent shows remembered for decades by those who witness them—like the Leonids of 1833 or, more recently, the Draconids* in 1933 or 1946, which follow in the wake of specific comets (comet Tempel for the Leonids, and Giacobini–Zinner for the Draconids). Other well-known showers, like the dependable Perseids (associated with the comet of Tuttle–Swift, 1862 III), which provide celestial fireworks during

* Each meteor shower is traditionally given the name of a constellation in which its 'radiant' happens to be situated.

several summer nights around the middle of August each year, are spread almost evenly along the entire orbit of the swarm. Very often, such swarms survive long after the parent comet has completely disintegrated and become only a memory of the past—like the comet Biela, last seen in 1852, which was the progenitor of the 1872 meteor shower of the still-active Andromedids.

Not all meteors intercepted by our Earth on its perpetual journey around the Sun belong to specific showers. Such showers cannot, in general, retain their cosmic identity much longer than the comets that gave rise to them; afterwards they disperse in the interplanetary substrate to become the meteors which we call sporadic. Although this sporadic reservoir of meteoritic material is vastly greater than that forming distinct showers, the latter exist locally in greater concentrations and encounters with them can produce more spectacular, albeit isolated, displays.

But one fact should be stressed above all: namely, that all meteors whose orbits have been determined with adequate precision—whether shower or sporadic—revolve around the Sun in closed elliptical orbits; none has been found to approach us along a hyperbola. This means that all meteors which we encounter belong to the solar system; none reaches us directly from interstellar space. An acquisition of interstellar debris which could be described as meteors is possible in principle (and such meteors could possess ages in excess of $4 \cdot 6 \times 10^{-9}$ years), but no single piece of such a meteorite is known to us so far; and even those which only flash overhead in the sky—if originally interstellar—have been thoroughly domesticated in the solar system before being picked up eventually by the Earth.

Zodiacal cloud

The meteors and meteorites encountered by the Earth in space are not the only solid matter we come across in the solar system. Another manifestation of its presence has long been known to us in the form of a circum-solar cloud producing the so-called zodiacal light. This is a cone of feeble light centred on the Sun, and extending away from it in the plane of the ecliptic (see Plate 14) among the constellations of the zodiac (hence the name). In fact, zodiacal light extends around the whole sky along the ecliptic as a zodiacal band, which decreases in brightness with angular distance from the Sun, but increases again slightly to form the so-called *Gegenschein* in an area several degrees in diameter near the anti-solar point, i.e. in the direction of the Sun–Earth axis projected beyond the Earth.

What is the source of this light? The fact that it appears to be roughly symmetrically disposed around the Sun—the two principal axes of

PLATE 14: Zodiacal light photographed by D. E. Blackwell at Chacaltaya in the Bolivian Andes (5200 m above sea-level) on 2 August 1958. Star trails indicate the duration of the exposure.

symmetry of the bulk of the zodiacal light being the ecliptic, or, rather, the invariable plane of the solar system (deviating by 1°43′ from the ecliptic) and the meridian planes passing through the Sun—discloses that the origin of this luminous phenomenon is sunlight, *scattered* on solid particles pervading the whole interplanetary space, and condensed strongly towards the plane in which the planets revolve. This has been confirmed by studies of the spectrum of the zodiacal light carried out in the late 1950s by Blackwell and his associates at Chacaltaya (5200 m above sea-level), and continued in the

1960s by other investigators from the substantially greater altitudes attainable by balloons and rockets.

The spectrum of the zodiacal light has proved, in effect, to be identical with the spectrum of the Sun, with all its spectral lines faithfully reproduced without essential change. The fact that these lines are not smeared out by the Doppler effect that could result from random motions of thermal origin discloses that the particles responsible for scattering are dust grains of microscopic size, and not (say) free electrons expelled by the Sun in the form of the solar wind. The density of this wind has, since the early 1960s, been repeatedly monitored by interplanetary spacecraft, and found to be too low to be of photometric significance. It may now be confidently affirmed that the zodiacal light is due almost wholly to the scattering of sunlight by dust-particles in interplanetary space; free electrons can play only a negligible role.

The measured distribution of intensity in the continuous spectrum of zodiacal light, and the degree of its polarization at different angular distances from the Sun, have shown that the average size of the particles of this interplanetary dust lies between 0·3 and 10 microns. Moreover, the refractive index of its particles discloses that, like the meteors, they consist predominantly of silicate material. This means that the mass of the average dust grain in the zodiacal cloud must be of the order of $10^{-14\pm6}$ kg. This is far below the limit of detection of meteoric material on the Earth by optical or radio methods. Photometry of the zodiacal light permits us, therefore, to extend our studies of the particulate contents of interstellar space to very much finer ingredients than those that we can detect as meteors or meteorites.

Photometry of the sunlight scattered by the zodiacal cloud indicates that the density of the cloud material diminishes (roughly) with the inverse cube of the distance from the Sun, and the mean temperature of its grains varies accordingly. For such grains not only scatter sunlight, but also (their albedo being very low) absorb most of it, and are thereby heated to temperatures which eventually lead to volatilization. These temperatures can actually be measured from the intensity of thermal emission of the zodiacal dust in the near infrared: such measurements were undertaken by Peterson and MacQueen in 1967–8. Their results have disclosed that, at a distance from the Sun equal to ten solar radii, the mean temperature of zodiacal cloud particles 0·1 μm in size is around 340 K. Particles 10 μm in size are heated to 1300 K, because larger grains find it more difficult to get rid of the heat which they absorb. This is close to the limit at which they are bound to volatilize. Indeed, below this limit silicate particles are rapidly destroyed by the heat of the Sun. Metallic particles can survive down to a distance of 3·5–4·0 solar radii; below this they are likewise completely destroyed by the solar furnace.

These facts, now attested by reliable observation, are not at variance with the theory that interplanetary dust is responsible for one aspect of the solar corona—a striking optical phenomenon which becomes visible to the human eye during total eclipses of the Sun. A part of the light of the solar corona (the so-called F-corona) proves again virtually identical with that of direct sunlight, with all its spectral absorption lines faithfully preserved, down to a distance of less than one solar radius from the apparent limb of the Sun. However, as was shown by Allen and van de Hulst as far back as 1946–7, this component of the solar corona originates by *diffraction* of sunlight on interplanetary dust between the Sun and the Earth, and not necessarily just in the proximity of the Sun. At low angular distance from the emitting body, diffraction becomes a much more effective process for redistribution of light than scattering. A familiar example of this is the diffraction rings around a distant lamp seen through fog, to which the solar F-corona represents a celestial analogy.

The fact that the zodiacal cloud is optically thin should permit us to study any systematic motions within it (such as may arise, for example, from its motion around the Sun) from the asymmetry of spectral lines of solar origin scattered from it in different directions in the sky. Although existing observations of this phenomenon are not yet conclusive, they indicate that the zodiacal cloud revolves around the Sun in the same direction as the planets, and with appropriate Keplerian angular velocity. In other words, the individual grains of the zodiacal cloud seem to revolve around the Sun as mass-points independent of each other, which means that their mean free path in space is probably very long in comparison with the dimensions of their orbits.

This fact is interesting, because it bears on the general stability of the zodiacal cloud and its age. The particles in the size range encountered in the zodiacal cloud are not acted upon by gravitational forces alone. Particles smaller than 0·3 μm in size would, in fact, be swiftly blown away from it by the radiation pressure of sunlight, and even particles appreciably larger than this limit, which can orbit around the Sun, are bound to experience a drag force due to the aberration of sunlight which would cause them eventually to spiral into the Sun and be destroyed.

It has been estimated (by Whipple and others) that a continuous operation of these processes means that the zodiacal cloud suffers a mass loss of the order of 10^4 kg per second. Since the total mass of the zodiacal cloud appears to be of the order of 10^{16} kg—comparable with the mass of a single comet rather than of a planet—the Poynting–Robertson effect alone could remove most of it in 10^{12} seconds, or 10^{5} years. These values are admittedly based on underestimates, and could be easily off by one, or even two, orders of magnitude. Even so, this time is so short in comparison with the probable age of the solar system that the zodiacal cloud we see today must represent only an ephemeral phenomenon, which could not exist without being continuously replenished.

The existence of numerous meteor streams with cometary associations suggests that comets may provide one steady source of material. This is, however, not likely to be the whole story; and the possibility should be kept in mind that some of it may also be accreted gradually from interstellar space. Such a process could not work directly, and extensive periods of domestication would again be required before the material in question could enter the inner precincts of the solar system and bask briefly (though for longer than the comets) in sunlight before its eventual destruction by the Sun, but whether or not this is really the case only the future can tell.

11
The Structure and Origin of the Solar System

IN CHAPTER 1 we gave a brief general outline of the principal types of astronomical population inhabiting our solar system, and in the subsequent chapters we have detailed their more individual properties. We wish now to cast a retrospective glance at what we know of our planetary family and other members of this system, in order to try to discern at least the general outlines of its design which may, in turn, hold a key to the secret of its origin.

The general structure of the solar system

First, let us consider the principal kinematic characteristics of our system, because such characteristics are well conserved (or change only exceedingly slowly) in the course of time. We have seen already in the preceding chapters that the motions around the Sun of most constituents of the solar system conform to a certain basic set of rules which can be briefly summarized as follows:

(*a*) their orbits deviate but little from circles;
(*b*) the planes of such orbits cluster around the invariable plane of the system (the orientation of which remains fixed in space), though their inclination to the solar equator is considerable; and
(*c*) the celestial bodies revolving in such orbits do so in the same direction.

The larger the mass of the body concerned, the more closely it follows these rules. Large planets observe them more closely than terrestrial ones such as Mercury or Pluto, let alone unruly little asteroids; and the comets show a flagrant disregard of them all. While larger bodies of the solar system seldom exhibit orbital eccentricities exceeding 0·1, cometary orbits can be almost arbitrarily elongated and exhibit eccentricities bordering on parabolic. In addition, the inclinations i of their orbits to the invariable plane of the solar system are distributed almost at random between 0° and 180°; those for which $90° < i < 180°$ revolve, in effect, in retrograde orbits.

It is easy to see what could have established such a set of traffic rules within the solar system, namely, the need for self-preservation. Whatever the initial disorder may have been in the new-born solar system, a compliance with the three rules above would ensure a maximum expectancy for survival; bodies which disobeyed them could perish in collisions, or be ejected from the system in hyperbolic trajectories as a result of perturbations caused by the major planets—as is still happening to the comets. If indeed the initial conditions with which our planetary system originated had been chaotic, many of its original constituents could have been lost to it by the dynamical action of their neighbours before the system assumed its present well-ordered form.

Such considerations should largely explain why most surviving members of the solar system revolve around the Sun in nearly circular orbits and in the same direction. That they do so also in almost the same plane probably goes back to the cumulative effects of planetary perturbations, mainly caused by Jupiter and Saturn, whose attraction gradually pulled the orbits of other lesser bodies to revolve in the same plane as themselves.

What is true of the planets applies by and large to their satellites. They too revolve around their respective central planets largely in the plane of their equators, in direct orbits deviating but little from circles, and conform to this pattern the more closely, the greater their mass or proximity to their central planet. They may do so for the same reasons as the planets themselves; and this would tend to suggest that the processes by which the solar and planetary systems originated were essentially the same and differed only in scale.

The present dimensions of the solar system and distribution of the semi-axes of the planetary orbits are almost invariant over long intervals of time, but not entirely so. Ernest W. Brown, who could speak to this subject with the greatest authority, in his last book (1933), put forward a view that the invariability of the planetary semi-axes can be trusted (on the basis of the available mathematical theory of planetary motions, based on asymptotic expansions which are eventually divergent) within time intervals of the order of a few hundred million years, but not more; and this is barely ten per cent of the age of the oldest rocks brought from the Moon, or of the oldest meteorites. In the course of the long past of the solar system, the absolute dimensions of planetary orbits, as well as their eccentricities and inclinations, may indeed have undergone considerable changes, but which way such changes may have gone is impossible to reconstruct.

Another dynamical characteristic of the solar system which we do not yet fully comprehend is the existence of certain striking *commensurabilities* between the orbital periods of many of its constituents. Thus, the orbital periods of Jupiter and Saturn, the largest planets of the solar system, are in the ratio of 0·4027, or almost 3 : 5; those of Uranus and Neptune are nearly 1 : 2; and the orbital period of Uranus is virtually 7 times as long as that of Jupiter, or one-third of that of Pluto. Such commensurabilities are not limited only to the planets. We find gaps in the distribution of asteroids in ratios commensurable with the mean motion of Jupiter (e.g. 2 : 1); and a clustering tendency a little off such exact commensurability (Kirkwood gaps). For certain ratios, such as 3 : 2 or 1 : 1, we find asteroids with almost exactly commensurable motions (see Figure 5).

A similar situation is also obvious in the systems of planetary satellites. Consider the Galilean satellites of Jupiter, where the orbital period 3·551d of its second satellite, Europa, is almost exactly twice as long as that of Io (1·769d), and one-half of that of Ganymede (7·155d); while the latter bears a ratio to the period of Callisto (16·69d) of very close to 3 : 7. In the system of Saturn, Cassini's division, the most conspicuous one of the several zones of avoidance for particles revolving in the ring, occurs at a distance from the planet at which an orbiting particle would orbit in a period of one-half of that of the innermost satellite, Mimas, or one-third of that of Enceladus, the second innermost satellite. Furthermore, Saturn's most massive satellite Titan revolves around its planet in a period of 15·95 days, which bears a ratio of 3 : 4 to that of Hyperion, or 1 : 5 to that of Japetus. The system of the five satellites of Uranus is characterized by any simple rational fractions between orbital periods of any two of its satellites, though simple relations appear to exist between integral multiples of mean motions of three or four of them.

The existence of such near resonances between orbital periods of several planets or their satellites constitutes a still unsolved problem of the celestial mechanics of the solar system, and their significance remains enigmatic. They appear to be too numerous to be the result of mere chance, and are probably connected with a continuing slow operation of dissipative forces within the system. Such forces are, however, bound to be extremely weak, and the fact that they seem to have succeeded in impressing appreciable effects on the structure of the system as a whole could be regarded as a dynamical testimony of its great age.

Another dynamical property which most members of the solar system (again with the exception of comets) have in common is their *axial rotation*, though the inclinations of the equators to their orbital planes appear to be rather erratic. For Jupiter, this inclination amounts to only 3°·1, and for the Earth (23°·5), Mars (25°·2), Saturn (26°·8), and Neptune (29°) it remains moderate. However, for Uranus the inclination attains 98° (so that the planet rotates in a direction opposite to that of its revolution around the Sun), and for Venus, the inclination is 177°, implying retrograde rotation (as if a forward-rotating system had been tipped over almost upside down!).

The *periods* of axial rotation (i.e. the durations of planetary solar days) are generally short for the major planets (and even more so for the asteroids), but very long for Mercury and Venus. Moreover, certain intriguing resonances have recently been found to exist between the rotation and revolution of the inner terrestrial planets. Thus, the sidereal day on Venus and its synodic revolution relative to the Earth appear to be almost exactly in the ratio 1 : 5, while the synodic day and year on Venus are in the ratio of 1 : 4, and on Mercury the durations of the solar and sidereal days appear to be in the ratio of 3 : 2.

Why this should be so also remains a mystery: the gravitational lock-in between the axial rotation of Venus and its relative position with respect to the Earth is particularly odd, for the tidal interaction between the two bodies at their present distance is extremely weak. In the case of our Moon (or of other satellites in the solar system) the prevalent tidal friction has been more than sufficient to synchronize the axial rotation of the satellite with its revolution about the central planet, but the real cause of the apparent coupling between the Earth and Venus—strongly suggested by recent radar data—remains still obscure.

Axial rotation and orbital revolution endows each planet with a certain amount of *angular momentum* (equal, basically, to the product of the mass and the angular velocity of its motion) which remains conserved even through long planetary lives and which, together with the mass itself, represents one of the two fundamental dynamical attributes of each body. When we evaluate the magnitudes of planetary angular momenta arising from their axial rotation and orbital motion we find that, *whereas the overwhelming bulk of the mass of the solar system rests in the Sun* (all the planets taken together do not add up to more than 0·13 per cent of it), *an almost equally overwhelming preponderence* (98 per cent) *of its total momentum is stored in the orbital momenta of the major planets*. This preponderance of the orbital momenta over rotational momenta among planetary bodies is analogous to the situation we encounter among the double stars in the sky, but at total variance with the state of affairs prevalent among the families of satellites of the major planets; for there the bulk, not only of the mass, but also of the total momentum of the system, belongs to the central body.

Composition of the solar system

In further pursuit of our inquiry into the past of the solar system we can scarcely do better than find out what this system has been made of. Turning our attention to the *chemical composition* of different classes of objects constituting the solar system, we find that the major planets, Jupiter, Saturn, and (to a lesser extent) Uranus and Neptune, are essentially of a composition chemically akin to that of the Sun, containing two-thirds to three-quarters of hydrogen by weight, the rest being largely helium with some admixture of elements of the carbon–nitrogen group for Uranus and Neptune. The masses of these planets range from 318 terrestrial masses for Jupiter to 14·7 terrestrial masses for Uranus.

When we come, however, to the Earth itself and to other terrestrial planets in the mass-range of 1·00–0·01⊕, we find something quite different: the principal chemical constituents of their interiors appear to be oxygen, silicon, and iron and the mean densities of their globes range between 4000 and 6000 kg m^{-3}, in comparison with 1000 ± 500 kg m^{-3} for the major planets. The asteroids probably belong to the terrestrial group. On the other hand, the comets with their hydrocarbons and other constituents are much more akin again to the major planets.

A contrast between the two groups of planetary bodies—those of solar and silicate composition—is indeed striking, and represents the most fundamental aspect of the problem of the origin and evolution of the solar system. The essential physical properties of the major planets have already been summarized in Table 2, and those of the terrestrial planets in Table 5. One can, however, ask if these present characteristics correspond to the original properties of these planets at the time of their birth? In order to attempt an answer, let us turn to the data compiled in Table 13, which lists the first eight elements occuring in greatest abundance in the interior of our Earth; column 3 gives the proportion by weight in which the elements contribute to the mass of the proto-Earth.

At the time of their origin, all planets—including the terrestrial ones—are thought to have condensed from primordial matter that we still find in the unadulterated state today in the solar atmosphere—for reasons which have already been mentioned on p. 4. And, last but not least, its chemical composition can be established, by quantitative spectroscopic analysis, in considerable detail.

Suppose that we take a sample of the gases now found in the solar atmosphere, and allow for the escape of its lighter constituents by relaxing the gravitational field: how large an amount of mass of solar atmospheric composition would be needed to produce, on condensation, a planet of the terrestrial mass and composition? The present abundances, in the solar atmosphere, of the nine elements listed in Table 13 are given in columns 4 and 5; the former gives the number of the respective atoms relative to

TABLE 13

Proportions by weight of elements in the earth's interior and the proto-earth

		Earth interior	*Solar atmosphere*		
Element	*Atomic weight*	*Percentage by weight*	*log N*	*Percentage by weight*	*Proto-earth (⊕ masses)*
Hydrogen	1·0		12·0		
Oxygen	16·0	27·2	8·96	1·46	(19)
Magnesium	24·3	11·3	7·40	0·061	190
Aluminium	27·0	1·07	6·20	0·0043	250
Silicon	28·0	13·8	7·50	0·089	160
Sulphur	32·1	2·74	7·30	0·064	43
Calcium	40·0	1·07	6·15	0·0057	190
Iron	55·8	38·8	6·57	0·021	(1870)
Nickel	58·7	2·7	5·91	0·0048	570

hydrogen on an arbitrary scale, while the latter gives their percentage contribution by weight. The last column lists the mass of the hypothetical proto-Earth consisting of the solar material which on escape of the more volatile component would leave us with a residue of approximately the amount of the respective element in the Earth as we know it today.

The mass of its hypothetical parent body based on the present abundance of these elements in the Sun and the Earth shows a considerable dispersion for different elements; that for oxygen is probably too low because of the high volatility of this gas, while that for iron seems much too high because of an apparent over-abundance of this element in the terrestrial interior. But if we disregard these two extremes, the average for the rest comes out to about 300 terrestrial masses for the original proto-Earth, and a quantity of the same order of magnitude for other terrestrial planets.

It is interesting to note that 300 ⊕ comes close to the actual mass of Jupiter, which consists of matter approximately the present composition of the solar interior. Are we, therefore, to conclude that Jupiter and the Earth (and other terrestrial planets) began as configurations of comparable mass and composition, and that their present differences in mass are due to the fact that while the Earth and the terrestrial planets have since lost most of their volatiles, Jupiter and other major planets managed to retain theirs? The answer may indeed be in the affirmative, and the possibility deserves further study.

What could have brought about such a difference between the terrestrial and major planets of the solar system in the course of their subsequent evolution? Part of the answer may be the general location of two groups of planets in space; for while all terrestrial planets (with the exception of Pluto) are confined to a region within 1 to 5 AU from the Sun, the distances of the major planets range between 5 and 30 AU. We do not know whether or not the absolute values of these distances have remained unchanged from the commencement of the existence of the solar system. But even if the individual planetary distances have changed appreciably in the course of the past 4·6 billion years, the relative distribution of distances may have remained the same. In particular, Jupiter may have always been more than five times as far away from the Sun as our Earth and, therefore, exposed to a solar climate some 160 K cooler than the one in which our Earth has evolved. Lower temperature would, in general, inhibit more effectively the escape of the volatiles; and the fact that the Earth has lost most of its initial endowment of these while Jupiter has managed to retain them may be partly due to this cause.

There may, to be sure, have been other contributory causes. For example, even if the initial masses of Jupiter and of the Earth were approximately the same, or at least comparable, the velocity of gas escape from their gravitational field would have been different because of the different size of their initial condensations. Suppose that the condensation of gas and solid particles from which the Earth originated was larger in size, or contracted more slowly, than proto-Jupiter. If so, this could have deprived our Earth of its volatiles at least as effectively as increased temperature. In fact, Jupiter and the rest of the major planets may have retained their original (solar) composition not only because their material was cooler, but also because they may have formed more compactly; or again because they collapsed with greater speed to their present size.

Such arguments make it at least plausible that the masses of all planets of the solar system may initially have been comparable, and that the present difference between major and terrestrial planets may go back to different rates of dissipation of their volatile matter at different distances from the

Sun (or, possibly, to their different original size or subsequent rate of compaction). However, let us now shift our attention to other aspects of the physical structure of the terrestrial planets with which we are more intimately acquainted, and which make them differ from each other within the group.

The first difference that we wish to single out concerns the *magnetic fields* of the terrestrial planets, or rather, the lack of them except in the case of the Earth. Our own planet possesses a very noticeable dipole magnetic field, with a surface strength of a sizeable fraction of one gauss. However, the space missions of the 1960s have disclosed that, among all terrestrial planets, the Earth is unique in this respect in the inner part of the solar system; for neither Venus, Mars, nor our Moon possesses any observable trace of external magnetic field.

How can we account for this difference? Although the process generating the terrestrial magnetic field is still far from being properly understood in detail, the dynamo theory of its origin lays down one prerequisite for its generation: an internal fluid core of high conductivity in rapid rotation. For the Earth, this condition is amply fulfilled (regardless of whether the terrestrial core consists of nickel-iron or silicate metallic phase, for both would be equally conductive). Mars rotates almost as fast as the Earth (in 24 hours and 37 minutes of our mean solar time), but possesses no heavy core to speak of and, as a consequence, its magnetic moment is less than $1–2 \times 10^{-6}$ of the terrestrial one, according to the results obtained by Mariners 6 and 7 in 1969.

On the other hand, Venus may well possess a core as large and massive as our own, but it rotates so slowly (its sidereal day is equal to 243 days of our terrestrial time, and the rotation is retrograde) as not to produce any magnetic dipole field of strength exceeding 0·002 of our terrestrial one. Our Moon likewise does not exhibit any such general field exceeding 10^{-5} of that of our own planet, because it does not possess any heavy core, and its rotation is slow (27·32 days). No direct observational evidence on the magnetic field of Mercury is as yet available; but its axial rotation time of 58·6 days may be too slow for excitation of the dynamo mechanism. Therefore, the magnetic properties of the terrestrial planets and differences between them can probably be understood in terms of the structure and angular momentum of their globes.

The same reasoning may apply also to Jupiter, the only other planet of the solar system now known to possess a magnetic field. The synchrotron emission by Jupiter in the domain of its radio spectrum indicates that this planet possesses a field of several hundred gauss—about 1000 times that of our Earth—probably generated by rapid rotation (with a period of less than ten hours) of a conducting core of solid (metallic) hydrogen.

When we turn our thoughts from planetary interiors to the surface, other intriguing differences confront us in the mass and composition of planetary atmospheres. The situation can be summarized briefly by saying that whereas the major planets are surrounded by massive and extensive atmospheres consisting predominantly of hydrogen and its compounds, those of the larger terrestrial planets (Earth, Venus, Mars) contain the elements and compounds of the carbon–nitrogen–oxygen group in very much smaller amounts; while still smaller bodies (Mercury, the Moon) possess no atmospheres at all. The main features of this situation can be understood in terms of the large mass-differences between different planets (together with different climates caused by different distances from the Sun). However, other differences are encountered—in particular, between the Earth and its adjacent celestial neighbours Venus and Mars—which are worth more than a passing notice. Some of these have been listed in Table 14. These figures are based (for Venus and Mars) on the results of recent space missions of the U.S. Mariners and the Russian Veneras.

The outstanding feature of these data is, first, a great difference in the total amount of gas surrounding Venus, Earth, and Mars. Whereas the air pressure on the Cytherean surface appears to be not less than one hundred terrestrial atmospheres, that on Mars is barely one-hundredth of an atmosphere. Most of the air around Venus is carbon dioxide—some 10^4 N m^{-2}, in contrast with only about 0·1 N m^{-2} of the same gas in the terrestrial atmosphere. To be sure, much of the terrestrial carbon dioxide is presently locked in solid carbonate compounds such as limestone, which are common in the crust of our planet. Geochemists have estimated that a total decomposition of all these carbonates could generate a carbon dioxide envelope

TABLE 14

Chemical composition of planetary atmospheres

	Venus	*Earth*	*Mars*
Air pressure	>100 atm	1 atm	0·0065 atm
Ground temperature	700 K	300 K	230 K
Nitrogen	<300 N m^{-2}	80 (200) N m^{-2}	$<0·1$ N m^{-2}
Oxygen	<1	20 (8×10^5)	$<0·01$
Carbon dioxide	$\sim10\,000$	0·08 (7 000)	0·6
Water vapour	100	40 (40 000)	$<0·001$

around the Earth of some 70 atmospheres air pressure; and if this is the case the total supply of carbon dioxide, both gaseous and fossil, on Venus and the Earth need not necessarily be very different; indeed, the sole difference may be the fact that whereas most part of the terrestrial carbon dioxide has been deposited in the solid carbonates of the Earth's crust, on Venus it has remained in gaseous form.

But why should there be such a difference? The clue is probably an equally gross disparity between the contents of liquid and gaseous water (see Table 14) on these planets. On the Earth, the global amount of atmospheric water vapour is (on the average) close to 4 N m^{-2}, and gives rise to a partial pressure of about 0·001 of an atmosphere. If, however, all the ocean water were to evaporate, the air pressure on the Earth would be increased 400-fold. On the other hand, in the Cytherean atmosphere water is 10 000 times scarcer than carbon dioxide; and this is, moreover, its total supply since, at a temperature of 400°C, no part of the surface of Venus can be covered with liquid water. The large amount of water on the Earth may explain why most of the terrestrial carbon dioxide has been locked up in solid carbonates, but why is Venus, like Mars, so excessively dry? And is the relatively large amount of gaseous carbon dioxide on Venus, or Mars, primordial, or was it liberated by gradual degassing of the interior? We do not yet know.

A similar discrepancy is encountered with atmospheric oxygen. On the Earth, it amounts to not less than 20 per cent of our air. Moreover geochemists tell us that at least 40 000 times as much is stored in different oxides of the terrestrial crust and much of this could also have been extracted in the past from the atmosphere. We do not know, of course, how much of such fossil oxygen may be contained in the crustal layers of our sister planets Venus or Mars; the well-known reddish colour of Mars may be indicative of the presence of iron oxides in its exposed surface. In their atmospheres, however, free oxygen now appears to be exceedingly scarce—much more so than water vapour itself.

Now oxygen is so reactive a gas that it cannot remain in contact with almost any kind of solid surface in the free state for an astronomically (or geologically) long time; sooner or later all of it is bound to be locked up in solid compounds. Geophysicists and biologists are in substantial agreement that our atmosphere can possess and maintain its 20 per cent content of free oxygen only because the supply lost through oxidation is constantly being replenished by the photosynthesis of green plants. In other words, most of the free oxygen we breathe is probably of organic origin; and since the amount of oxygen on Venus or Mars appears to be negligible, this can be taken as almost sufficient proof of the fact that no life on appreciable scale can exist there at the present time—although this is not, perhaps, a surprising conclusion.

Lastly, we shall mention another point of general interest, connected with the observed abundances of heavy radioactive elements in planetary crusts. It has been known for a long time that the observed uranium or thorium content of rocks (in particular, granites) constituting the Earth's crust must be much higher than in the deeper parts of the mantle or in the core, for otherwise radiogenic heat produced by the disintegration of such elements would have been sufficient to melt the Earth completely in the course of its long geological past.

Since 1969, however, the results of the Apollo missions have brought to light the fact that the relative abundance of uranium and thorium in lunar rocks is even higher than on the Earth. Since, moreover, the lunar crust is very much more rigid than the terrestrial mantle, chemical differentiation processes, sometimes invoked for concentrating radioactive elements in the upper parts of the mantle, cannot readily be applied to the Moon. A relatively high concentration of uranium and thorium would therefore represent an even harder cosmochemical puzzle than has been the case on the Earth.

A clue to its possible solution may be provided by the extreme age of lunar rocks, many of which solidified as long as 4·6 billion years ago. At that time, as we pointed out in Chapter 7, our Sun may have still been in the last throes of its contraction, and its surface would not only have been nearer to us, but its powerful solar wind, some 10^7 times more intense than at the present time, would have contained an appreciable fraction of neutrons. An absorption of these neutrons by surface material of planets in the last stage of their formation may have produced a short spell of nucleogenesis, of which the anomalously high abundance of uranium and thorium and certain other heavy nuclides may represent a surviving remnant.

If this was the case, then we should expect the surface of Mercury to be even more radioactive than that of the Moon, while the surface material of Mars should be less radioactive than that of the Earth or the Moon. Space missions to Mercury and Mars in the 1970s may confirm or disprove these expectations. But until they have delivered their verdict, the possibility that the young Sun may have sprinkled the surfaces of new-born planets with a neutron gun before large-scale convection died out in its interior should be kept in mind.

Origin and evolution of the solar system

Having briefly surveyed the principal dynamical features of the solar system and its chemical composition, let us now turn to the grand design of this system and inquire about the *origin* of the system as a whole. While we are still quite far from being able to reconstruct the details of such a process with any assurance, certain general guidelines which are unlikely to be misleading have already emerged.

First, we are now virtually certain that—contrary to the view held from the days of Kant and Laplace up to almost the middle of this century—the formation of our planetary system represented an act *concurrent* with the formation of the Sun, rather than consecutive to it. Second, we are now reasonably sure *when* it occurred, from the time which has elapsed since the solidification of the oldest rocks or meteorites now in our hands. We mentioned in Chapter 10 that the ages of meteoritic material that our Earth keeps picking up on its journey through space have been found by radioactive methods to cluster around $4{\cdot}6 \times 10^9$ years, and extensive recent determinations of the ages of lunar rocks brought to the Earth by the Apollo missions (Chapter 7) came very close to the same value—in fact, to a mean of some $(4{\cdot}63 \pm 0{\cdot}01) \times 10^9$ years.

Now we stressed earlier the fact that ages so determined refer to the time elapsed since the respective samples solidified. It is, moreover, certain that rocks on the Moon, or meteorites in space, solidified in places far removed from each other. If such random samples are all of the same age, it follows that they were all formed as a part of the general process of cooling of the entire solar system; and this, therefore, seems to have occurred some $4{\cdot}63 \times 10^9$ years ago. Moreover, a very small dispersion of the determination of ages clustering around this limit indicated that this process, once it started, did not last very long; probably less than 100 million years.

Furthermore, several phenomena recorded in the lunar crust, mentioned in Chapter 7, lead us to believe that when the bulk of the lunar surface solidified, the Sun was still in the last throes of its contraction—i.e. was only approaching the end of its own birth pangs. Besides, we know of nothing that could have induced the Sun to give birth to a planetary system. It is true that most (98 per cent) of the angular momentum of the solar system as a whole resides in the orbital momenta of its major planets, and not in the axial rotation of its central star. If, however, we were to transfer all planetary momenta on to the Sun, its axial rotation would speed up from 27 days to about 12 hours. So fast a rotation should render the Sun very appreciably oblate, but this oblateness would be still far from the limit at which the Sun would become equatorially unstable. Besides, the solar equator is inclined by 7° to the invariable plane of the planetary system, so that even if the Sun had ever shed any mass off its equator to form the planets, it would have done so in the wrong direction!

A tidal disruption of the existing Sun in the course of a close approach of another star—a process proposed by Jeans and Jeffreys in the first half of the twentieth century—now also seems to be out of the question. The density of stars in space is known to be so low that a close encounter of any two of them at a distance that would lead to tidal disruption could occur, at most, only a few times in the galaxy during its entire life-span. If so, planetary systems should be excessively rare in the Universe; and yet we now know that at least 1 per cent (and probably more) of the stars in our neighbourhood possess companions whose mass is of planetary, rather than stellar, order of magnitude. Therefore, planetary systems would seem to be reasonably common, not rare, in the realm of the stars that we know best; and if so, their origin cannot have anything to do with close stellar encounters.

Besides, it is virtually certain that no material now constituting the planets could have come directly out of the interior of the Sun by any process. Such gas would have been so hot as to dissipate thermally in a few hours (Spitzer, 1939) without a ghost of a chance of condensing into planetary globes. No, the formation of the planets must have been the result of some process that went on in parallel with the formation of the Sun itself—i.e. through the collapse of a primordial cloud of material which may have contained not only solid particles and neutral gas, but also plasma. In the presence of a magnetic field such a collapse may have represented, not only a mechanical or hydrodynamical but also a hydromagnetic problem (Alfvén and Arrhenius, 1970). Moreover, the total mass that went into the formation of the planetary companions of the Sun at that stage may well have been of the order of $0{\cdot}01\odot$, of which about $0{\cdot}001\odot$ is still left around today. While our present major planets may have retained the bulk of their original mass throughout their entire past, the terrestrial planets may represent only the residues of masses they may have once possessed before the loss of their more volatile elements.

If spherical symmetry could have been retained throughout this process, the result would have been the creation of a single star. We know, however, that in many instances stars do originate in pairs; and if the collapse occurs towards a *plane* rather than a single centre (perhaps as a result of an excess of angular momentum), the intermediary stage would be a disc which might well eventually condense into a planetary system. The exact mechanism of the terminal stages of such a process may as yet be debatable, but if planetary systems exist, Nature must be able to accomplish this task even if she has not yet seen fit to tell us how. But one aspect of the problem seems clear: the parent substrate from which the planets of the solar system

—including our Earth—originated must have been cold, or only moderately warm. It could, at some time, have been heated up to a temperature of a few thousand degrees, but not tens of thousands; for otherwise this material would not assemble into individual globes of planetary masses and compositions.

If we now regard our solar system as having been born at the same time as the Sun, and from a material similar to that which now constitutes the bulk of the Sun's mass, we can do so with perhaps one reservation; and that concerns the as yet doubtful origin of the *comets*. Did these too originate as a part of the same creative process, or are they instead the products of interstellar space which we pick up only on our orbital journey around the galactic centre? It is clear from their high abundance as well as their relatively rapid perishability that there must be a source from which their supply is being constantly replenished if we are to have any of them around us at all. Whether this course is to be sought in a cometary cloud at a distant periphery of our system (Oort, 1950) which we drag along with us through space, or whether we accrete them directly from interstellar space (Lyttleton, 1948) remains as yet obscure.

To be sure, none of the comets (or, for that matter, meteors) that we have a chance to observe from the Earth comes to us directly from interstellar space (i.e. approaches us along a hyperbolic path). Reserves drawn from interstellar supply may, however, have undergone a prior extensive process of domestication in the outer parts of the solar system before some of them accidentally penetrate into its inner precincts to become known to us, but their aboriginal birthplace or domicile continues to be shrouded in a mystery deeper than that which still surrounds some of the circumstances of the origin of other, more legitimate, members of our system.

But after this aside about the comets let us return again to the planetary globes and attempt to trace their evolution across the gap separating us from the time of the origin of the solar system. We know from several different lines of independent evidence, the lunar surface and meteorites for example, that at least a large part of the primordial material from which the solar system originated solidified within a relatively short time under highly *reducing* conditions, which were rich in hydrogen but extremely deficient in free oxygen. Therefore, the atmospheres of the major planets, with their tremendous amount of hydrogen and hydrogen compounds, may well be primordial or largely so, but those of the terrestrial planets were probably created by degassing of their interiors as a result of a gradual build-up of internal heat.

For even if the Earth and other terrestrial planets akin to it accumulated in the cold state, such traces of long-lived radioactive elements (uranium-235 and -238, thorium-228, potassium-40) as may have been present in their parent material commenced their slow work of heat production which, in due course, could have raised the internal temperature of such planets from several hundred to a few thousand degrees. Internal heating of planetary interiors by radiogenic heat released by the elements mentioned above is a very slow process, whose efficiency should (for a given concentration of the radioactive elements) depend essentially on the ratio of the volume throughout which heat is produced to the surface through which heat can be lost to space. Therefore, larger terrestrial planets like the Earth or Venus could have, in time, accumulated and stored more internal heat than Mars, let alone Mercury or the Moon, and such facts as we know about them seem to bear this out in a rather convincing manner.

As a by-product of this gradual build-up of radiogenic heat, planets like the Earth or Venus could, in time, have expelled enough volatile elements or compounds from their interiors to surround themselves with their present atmospheres or hydrospheres. Of course, we know nothing directly about the evolution of the Cytherean outer layers in the course of time, and nothing about our Earth during the first thousand million years (the dark aeon) of its existence. However, since about $3{\cdot}5 \times 10^9$ years ago the thread of the terrestrial evidence (in the form of rocks extant from those days) has become almost uninterrupted. It discloses that our Earth then already possessed an atmosphere whose main constituent was nitrogen (probably liberated by de-gassing of the interior) with some (possibly primordial) argon, but very little hydrogen or helium. The oxygen content was probably very much less than it is today; its gradual increase due to photosynthesis of green plants has brought about the only major change in composition of the terrestrial atmosphere since that time.

Not only was the air around the Earth some 3·5 billion years ago not unlike that we breathe now, but liquid water was present as well, though the seas of those times may have lacked their present salinity. But the fact that the oldest rocks preserved from that epoch are metamorphosed sediments discloses that water was already then on the Earth's surface in *liquid* form—which means that its mean temperature must have been above freezing point.

This fact is very interesting, for if we unwind the evolution of the present Sun for 3·5 billion years into the past we find that its luminosity should then have been 50–60 per cent *smaller* then than it is today. If the present radiation of the Sun is capable of maintaining the surface of the Earth at a mean temperature of some +15°C, the Sun of $3{\cdot}5 \times 10^9$ years B.C. would have been barely capable of maintaining it at $-10°C$, or definitely below the freezing point of water. If, therefore, sedimentary rocks of that age testify to the presence of liquid water on the Earth, it follows that the Earth must have been either at least 10 per cent nearer to the Sun than it is today—which is possible—or the greenhouse effect of the atmosphere at that time must have stored sunlight more effectively (perhaps because of a higher abundance of carbon dioxide). Neither possibility can be excluded, and both may have cooperated to bring about the moderate climate which has enabled water to remain in the liquid state ever since.

The subsequent fortunes of the terrestrial crust, with its consumption and drift of continental land masses and ocean floors—and, above all, with the emergence of life—are wholly beyond the scope of this book. We must, instead, close the present chapter with a glance at the more distant cosmic future of the solar system and, above all, of our Earth. What does the cosmic march of time hold in store for us? What, in particular, will be the future cosmic climate to which our planet is likely to be exposed?

As far as the internal sources are concerned, the gradual warming-up of the Earth's interior is likely to continue at least for another $1–2 \times 10^9$ years (depending on the more exact proportions of long-lived radioactive elements whose half-lives range from $0{\cdot}713 \times 10^9$ years for uranium-235 to $13{\cdot}9 \times 10^9$ for thorium-232), but scarcely much more; and the present growth of the fluid core at the expense of the mantle should be almost at an end. Not much more juvenile water from the interior can be expected to augment our oceans by continuing desiccation of the mantle, nor will our atmosphere be enriched much by its degassing in the future.

This means that the surface climate of our planet will, in the future, be increasingly at the mercy of the heat received through sunlight, and this fact bodes no good for our future. For with its secularly diminishing store of hydrogen fuel in the deep interior, our Sun is bound to evolve to a star of increasingly higher temperature—a trend which must eventually render the temperature of the Earth so high that our oceans will evaporate, and our atmosphere disperse to expose the naked surface of our planet to scorching heat in which nothing could survive—like on Mercury today. Thus the cosmic episode which commenced some $4{\cdot}5 \times 10^9$ years ago with the formation of the solar system will find its end in a heat-death under the scorching breath of the merciless Sun smarting under the increasing scarcity of hydrogen in its deep interior. Nothing that we or the planets can do will rescue us from its ultimate fiery embrace.

The only cheerful note here is the assurance that this dismal end is still a long time in the future. Very probably another 5–7 thousand million years or even more are in store for our descendants to enjoy sunshine before it gradually changes from a blissful friend to a relentless enemy. This trend of events may be accelerated if, in the course of time, the planets move closer to the Sun to meet their doom, or delayed if the planets tend to spiral slowly outwards. But when the end comes, the story of the solar system—from the oldest rocks on the Moon and the first living matter in the shallow waters of the Earth to the most exalted heights which humanity may yet attain in the future—will probably have lasted not less than 10^{10} years. Surely that should be long enough to have made this particular cosmic episode thoroughly worth while.

12
Other Planetary Systems in the Universe

IN THE PRECEDING chapters we have been concerned with one planetary system in the Universe: the one surrounding our Sun, of which the Earth is only one not-too-conspicuous member. It is the only planetary system with which we are acquainted with any degree of intimacy, but it need not be the only one that exists in the Universe around us. Until quite recently, we just did not know; and our ancestors could only guess at the scarcity or plurality of planetary systems among the stars, depending on their particular predilections or idiosyncracies.

It is only in recent years that advancces in precise astrometry have told us of the presence of celestial bodies, in our immediate stellar proximity, with masses of planetary rather than stellar order of magnitude; and their existence as well as the numbers in which they occur, have thrown a dramatically new light on the problem of planetary systems in the Universe.

Although the search for planetary bodies in our neighbourhood has only recently been successful, a quest for planets of other Suns has been with us for a very long time. The principal obstacle to the discovery of planets in the proximity of their central stars is, of course, the overwhelming disparity in brightness of these two classes of celestial objects, and a combination of this disparity with close proximity makes any observational discovery of the planets utterly impossible with any kind of optical means available to astronomers on the surface of the Earth.

In order to appreciate the obstacles in the way of such a discovery, consider the chances of discovering the existence of Jupiter, let alone of the Earth, from the distance of our nearby stellar neighbours in space only a few parsecs away. Even from the distance of α Centauri, our nearest celestial neighbour 4·28 light years distant, our Sun would appear in the sky as a star as bright as α Centauri appears to us ($+0^{m}\cdot 3$ apparent visual magnitude); while Jupiter, which reflects only $1 \cdot 5 \times 10^{-9}$ of sunlight falling on it, would be a starlet of $22^{m} \cdot 8$ apparent magnitude. A star so faint would be on the threshold of detection of the 200-inch telescope of Mount Palomar even if it had no bright stellar neighbour; but with the Sun at most only $3'' \cdot 96$ away, its chances of discovery would be virtually nil. Chances would be absolutely nil for the discovery of the Earth, an object four times fainter (Jupiter being ten times as large as the Earth, but five times as far as the Earth from the Sun), whose elongation from the Sun would never exceed the parallax of the star which for α Centauri is equal to $0'' \cdot 76$.

Suppose, however, that from some favoured direction in the sky Jupiter could project itself against the disc of the Sun and transit across it, as Mercury or Venus is occasionally seen to do from the Earth. Since the radius of Jupiter is one-tenth of that of the Sun, its transit across the Sun would diminish the apparent brightness of the Sun by one per cent or 0·01 of a magnitude—a change lasting some 40 days every 12 years if the transit were central, and dwindling to nought if the transit was as much as $0^{\circ} \cdot 2$ off centre for an external observer. The chances that such a phenomenon could be detected, or that we could spot planets attending other stars in this way, is, while not zero, extremely small; and certainly none has yet been discovered in this way.

Another more indirect method for detection of planetary systems in close proximity to the stars was proposed in 1961 by the Russian astronomer Fessenkov. It is based on a search for the zodiacal light (Chapter 10) around a star which is likely to accompany each planetary system, for its particles may represent left-overs from the time of the system's formation. The total amount of sunlight scattered by our own zodiacal cloud is several times greater than the light of all planets, so that a disparity in brightness between the central star and the object sought after around it would not be so overwhelming. Moreover, the direction in which a zodiacal cloud around other stars would be elongated could indicate to us the orientation of the invariable plane in which the principal planets revolve, and thus the locus in which their optical images should be sought.

On the other hand, although the total brightness of such a cloud is likely to exceed that of all individual planets, its surface brightness may again prove too low for observational detection; and the need to separate it from the scattered or diffracted light of the central star might well turn out to be as difficult as the detection of individual stellar images. At any rate, no discovery of any such phenomenon has so far been reported and none may be forthcoming while we are condemned to observe the skies only from the surface of the Earth through our atmosphere.

As long as this is so, and observational difficulties make it impossible for us to search with success for close companions of nearby stars in their own light, their planetary companions shining by reflected light can be detected with any hope of success only by their *gravitational* effect—or, more specifically, by the effects which the gravitational attraction of even a small mass can exert on the space motion of its more substantial partner. A single star is bound to travel through space with an essentially uniform velocity, influenced by other cosmic masses (such as the galactic arms or nucleus) to a negligible extent over intervals of time ranging from decades to centuries; and the projections of such uniform motions on the celestial sphere (the proper motions of single stars) should be rectilinear. On the other hand, if the star possesses a companion whose presence compels it to revolve around a common centre of gravity, its apparent proper motion in the sky will cease to be rectilinear, and will exhibit periodic oscillations around a straight line with a period identical with that of the orbital motion; the amplitude of the oscillation offers a clue to the mass-ratio of the system.

A careful search for the existence of such 'astrometric' binaries in our stellar neighbourhood has led in recent years to the discovery of several systems with companions which are invisible in their own light, and whose mass is less than one-hundredth of that of our Sun—i.e. of planetary rather than stellar order of magnitude. One example of such an object is the well-known Barnard's star, notable for its large proper motion and parallax. The star itself is a red dwarf of spectral type M5 and apparent visual magnitude 9·5. Its annual parallax of $0''\cdot545$ renders Barnard's star the nearest known star to us with the exception of the system of α Centauri, separated from us by a mere 5·98 light years. But what makes Barnard's star outstanding among all others is its large proper motion in the sky of $10''\cdot3$ per annum (which means that it moves through a distance equal to the angular diameter of the Sun or the Moon every 180 years)—the largest of any star known. Continued tracking of the proper motion of this star across the sky by P. van de Kamp and his associates has shown that this motion is not strictly rectilinear, but shows periodic fluctuations in direction with a period of close to 25 years, and an amplitude indicative of the presence of an invisible companion whose mass has been estimated by van de Kamp to be $0\cdot0015\odot$, or $1\frac{1}{2}$ times that of Jupiter!

Barnard's star is, moreover, not the only one of our stellar neighbours whose proper motion across the sky has disclosed the presence of unseen companions of small mass. Thus the brighter component of the visual double star 61 Cygni—the star whose parallax of $0''\cdot292$ (corresponding to a distance of 11·1 light years) was established by Bessel in 1837—has more recently been shown by Strand to exhibit wiggles in its proper motions with a period of only 4·8 years, indicative of the presence of a companion of mass 0·008☉. Moreover, another nearby star, Lalande 21185, our third-nearest stellar neighbour in space at a distance of 8·2 light years, was shown by similar methods to possess a companion of mass of the order of 0·01☉ revolving around the central star (a red dwarf of spectral class M2) in a period close to 8 years.

These are the three most obvious pieces of evidence for the existence of celestial bodies with masses of planetary rather than stellar order of magnitude in our stellar neighbourhood. All three are no more than twelve light years away, at a distance within which we find a total of seventeen stars including our Sun. If, therefore, four out of seventeen stars in our close neighbourhood appear to possess planetary companions of masses comparable with Jupiter (and less massive companions could easily escape detection), planetary systems may be possessed by some 25 per cent of all stars—possibly more.

In other words, this evidence tends to indicate that planetary systems, far from being rare, are in fact very common in the Universe. Our Galaxy alone is known to consist of some 10^{12} individual stars; and if even one-tenth of these were to possess planetary companions, the total number of planetary systems in the Galaxy would evidently be enormous!

If such is indeed the case, the question may arise as to the number of planetary systems which may be formed around us. In the preceding chapter we estimated that the formation of the solar system may have taken some 10^7 years, while its duration may be limited to 10^{10} years. It follows that even if only one-tenth of all stars possess planetary systems, one should be in the process of formation at any time among 10^4 stars. While the presence of planets of small mass may be difficult to detect for any but the nearest stars, their formation—while it lasts—may be attended by phenomena which are very much more conspicuous.

Is there any evidence among the stars around us that planetary systems akin to our own may be in the making at the present time? The answer may well be in the affirmative; and, in particular, the giant eclipsing system of ϵ Aurigae may be exhibiting symptoms of such a formative process at the present time. But whether or not this is actually the case remains as yet to be settled by future investigations.

13
Epilogue—Planetary Research in the Future

IN CHAPTER 1 we stressed the fact that most of our knowledge of the solar system has been acquired so recently that a book on this subject written only ten years ago would read like Latin or Greek in comparison; and the reader who has persevered so far can judge for himself the validity of this contention. It applies not only to the wealth of new observational data secured by a more direct contact with the objects of our study by means of radar or spacecraft, but also to the studies of the motion of the celestial bodies in the solar system with the aid of modern computing machines.

From the beginning of the eighteenth century to the middle of the twentieth, the solar system has been the proving-ground for the development of celestial mechanics. The possibility of making accurate predictions of planetary positions on the basis of Newton's law stimulated, on one hand, a gradual development of precise techniques for measurements of the positions of heavenly objects on the celestial sphere and, on the other, the development of computations exact enough to enable us to compare accurate observations with increasingly refined theory.

Dynamical astronomy of the solar system was the first branch of modern science to be based on natural laws which are as simple as they are exact; and its development into the magnificent edifice of celestial mechanics by Newton, Euler, Lagrange, Laplace, and Gauss, to name only the greatest, influenced profoundly the whole scientific outlook of mankind. The latter part of the nineteenth century—the period from Gauss to Poincaré—saw some splendid theoretical advances which proved, however, to be of increasingly limited practical use.

The development of positional measurements has gradually drifted to a stage at which the law of diminishing returns has become operative. The traditional methods in celestial mechanics for comparing theory with observations have been based on the measurement of the right-ascension* and declination* of a celestial object, in equatorial coordinates, at a given time. Through the cumulative efforts of several generations methods have been worked out which have enabled us to measure such coordinates individually within errors of the order of one part in ten million of a quadrant—but not more, mainly because of atmospheric refraction anomalies affecting the precision of the measurements in declination. The accuracy of measured right-ascensions is again limited by a finite precision of our astronomical clocks, be they the rotation of the Earth, or various laboratory devices.

The second half of this century has brought about a complete change in this time-honoured picture, with a far-reaching gain in accuracy being attained by an application of novel techniques both in theory and observation. In the first half of the twentieth century, astronomical determinations of position were still among the most precise measurements to be made in any branch of physical science. However, as a result of a technological spin-off from the Second World War, we learned to measure short intervals of time (or, what is equivalent, the frequency of periodic phenomena) with an accuracy 100 or even 1000 times higher than previously, and dynamical astronomy of the solar system became one of the first beneficiaries of such advances.

As a result, we no longer determine the position of a planet by measuring its right-ascension (α) and declination (δ) on the celestial sphere, or its distance by triangulation of a parallax; we measure its distance r and radial velocity $\dot{r}$ of approach or recession by radar—the distance by timing the return of reflected radar echoes and its velocity by measuring the Doppler shift of the returning signals. The range of this method is, of course, limited to distances from which measurable echoes can be returned. But it has already been used over the entire inner part of the solar system, from Mercury to Mars, and contacts are being established with Jupiter. The distances measured in this way are no longer expressed in kilometres or any other direct units of distance, but rather in light seconds or minutes—measured to a precision considerably greater than that to which we presently know the value of the velocity of light.

As a result of radar contacts already established with the inner planets, the dimensions of the solar system have been redetermined by microwave radio techniques with a precision more than a hundred times as high as that attained previously by astronomical triangulation, and a transition from radar to laser links (established so far only in the Earth–Moon system) should enable us eventually to gain in precision by another factor of 100—a truly fantastic accuracy, completely undreamt of only one generation ago!

Needless to say, such far-reaching advances in tracking of the motions of celestial bodies in the solar system have stepped up corresponding demands on the accuracy of the theory needed to interpret such motions. The time-honoured analytical procedures providing the link between the observed data and the time, based on series expansions, are no longer adequate for the purpose—especially as the type of the input data (i.e. r and $\dot{r}$ in place of α and δ) has changed completely with the advent of the new observing techniques. Theories rigorous enough to exhaust the precision of the new types of data can no longer be presented conveniently in the form of analytical expressions suitable for publication in print. They exist so far only in the form of machine programmes intelligible to automatic computers which alone can furnish the results of requisite precision in numerical form.

Thus, while the place of astronomical observers for studies of the distances in the solar system has, during our lifetime, been taken over by radio engineers, future books on celestial mechanics are going to be written, not for the general public, but for the automatic computers to which they will be primarily addressed. The sacred precincts of celestial mechanics will henceforward be restricted further to admit only those mathematicians who can converse with the giant computers.

The significant advances of this new astronomy are indeed of very recent date. The first radar contact was established with the Moon in January 1946, and the laser link (by Surveyor 7) in January 1968. The first radar link with a planet (Venus) came in March 1961; and for Mars, two years later. In the physical exploration of lunar and planetary environment, the pioneers have been the early Russian Lunas to the Moon (in September and October 1959), the second of which unveiled for us a glimpse of its far side, and the American Rangers 7–9 in 1964–5 which gave us the first glimpse of the lunar surface on the scale of one metre.

The real breakthrough began, however, with lunar soft-landers of both American and Russian origin in 1966–8, which furnished the first information on the physical and chemical composition of the lunar surface. These were followed since 1969 by the Apollo manned missions whose success will remain for ever a pride and glory of this turbulent century. For since July 1969—in fact, since Christmas Eve of 1968 when three American astronauts of the Apollo 8 mission became gravitationally attached to the Moon for almost 20 hours—*Homo sapiens* left its terrestrial cradle and took the first tentative steps towards its proliferation throughout the solar system.

The planets Venus and Mars—our nearest celestial neighbours after the

Moon—became targets of fly-by missions in 1962 with the U.S. Mariner 2 to Venus on 14 December of that year, and Mariner 4 to Mars on 14 July 1965; these were followed by several other probes of American and Russian origin, which are listed in Table 15. Some of these (Venus 5 to 7) descended softly through the Cytherean atmosphere by means of parachutes.

In the 1970s we shall witness many further exploits by spacecraft which will considerably widen the scope of planetary exploration. These include not only soft-landers on the solid surface of Venus, but also orbiters and soft-landers around Mars. Mercury should be the target of a fly-by mission

TABLE 15

Planetary probes launched in 1961–71

Name	*Origin*	*Date of launch*	*Destination*
Venus 1	U.S.S.R.	12 February 1961	Venus (contact lost after 7·5 million km)
Mariner 2	U.S.A.	26 August 1962	Venus (fly-by on 14 December 1964)
Mars 1	U.S.S.R.	1 November 1962	Mars (contact lost after 105 million km)
Zond 1	U.S.S.R.	2 April 1964	Venus (contact lost en route)
Mariner 3	U.S.A.	5 November 1964	Mars (failed by shroud malfunction)
Mariner 4	U.S.A.	28 November 1964	Mars (fly-by on 14 July 1965)
Zond 2	U.S.S.R.	30 November 1964	Mars (batteries failed after 5 May 1965)
Venus 2	U.S.S.R.	12 November 1965	Venus (fly-by on 27 February 1966)
Venus 3	U.S.S.R.	16 November 1965	Venus (crash-landing on 1 March 1966)
Venus 4	U.S.S.R.	12 June 1967	Venus (parachute landing on 18 October 1967)
Mariner 5	U.S.A.	14 June 1967	Venus (fly-by on 19 October 1967)
Venus 5	U.S.S.R.	5 January 1969	Venus (parachute landing on 16 May 1969)
Venus 6	U.S.S.R.	10 January 1969	Venus (parachute landing on 17 May 1969)
Mariner 6	U.S.A.	24 February 1969	Mars (fly-by on 31 July 1969)
Mariner 7	U.S.A.	27 March 1969	Mars (fly-by on 5 August 1969)
Venus 7	U.S.S.R.	17 August 1970	Venus (parachute landing on 15 December 1970)
Mars 2	U.S.S.R.	19 May 1971	Mars (ejection of capsule on 27 November 1971)
Mars 3	U.S.S.R.	28 May 1971	Mars (soft-landing capsule on 2 December 1971)
Mariner 9	U.S.A.	30 May 1971	Mars (orbiter since 13 November 1971)

in 1973, and the first Pioneer rocket was already launched in 1972 towards Jupiter to serve as forerunners of a grand tour of the solar system by more substantial spacecraft, making close calls on Jupiter, Saturn, and possibly also Uranus, Neptune, and Pluto.

The timing of this grand tour has been chosen for us by celestial mechanics and planetary configurations obtaining at that time. Between 1979 and 1990, the positions of Jupiter, Saturn, Uranus, Neptune, and Pluto in their orbits will be such that a single spacecraft launched from the Earth could swing around Jupiter, and then Saturn, to head for Uranus and Neptune before eventually escaping from the solar system. Planetary configurations to make this possible will not recur again for another 179 years.

A conventional spacecraft sent to Pluto by direct course would have to spend almost 40 years *en route* before reaching the proximity of its target. Planetary configurations between 1980 and 1990 should enable us not only to do all the outer planets as a part of the same mission, but to do so on a reduced time-scale—by taking advantage of the accelerations experienced during fly-bys of Jupiter and Saturn—without any extra expenditure of fuel. In such a way, the travel-time from the Earth to Pluto could actually be reduced from some forty to only eight years! We may even see this within our lifetime.

Moreover, looking further into our crystal ball, we anticipate that in the mid-1980s sufficient progress will have been made for man to land on the surface of Mars—the second (and last) planetary body on which man can land and return with impunity.

All these recent developments, and many more yet to come, signify that in the second half of this century the exploration of the solar system has ceased to be the primary concern of astronomy alone, and its problems and results will be increasingly shared between several branches of science and engineering. Astronomy has in the past been debarred from the status of a genuine experimental science by the utter remoteness of the objects of its study. With the exception of meteors—those small freaks of interplanetary matter intercepted by the Earth on its perpetual journey through space—the properties of all celestial bodies could only be studied by our telescopes at a distance, from the effects of attraction exerted by their masses, and from the ciphered messages of their light brought to us by photons across the intervening gaps of space.

The dramatic emergence in the 1960s of spacecraft capable of disengaging themselves from the gravitational field of our Earth and paying close calls on our planetary neighbours brought about a complete change of scene. Telescopes, and other traditional means of astronomical observations, will retain their use for studies of transient phenomena visible on planetary discs, but exploration *in situ* will increasingly become the task of non-astronomical instrumentation. This is as it should be, for as we extend the scope of our cosmic domicile from the Earth to our celestial neighbours in the solar system, the exploration of this domicile and its eventual utilization by the human race will become an issue of much wider concern to our descendants than any single academic discipline.

Not many men may reach the Moon, let alone Mars, within our lifetime. Yet all the rest of us, onlookers in this great episode of human adventure, should rejoice at the thought that, after centuries of slow growth and gestation, three distinct branches of human technology—rocket propulsion, radio communications, and computer control—matured sufficiently for their cooperative effort to make space travel and all that it brings with it an accomplished fact. That this came to pass within our lifetime is a historical accident for which we—the privileged generation—should be duly grateful to our fate.

Suggestions for Collateral Reading

ASTRONOMY OF THE solar system constitutes a subject whose roots go back many centuries, and the best single general source covering its early evolution still remains J. L. E. Dreyer's treatise on the *History of the Solar System from Thales to Kepler* (Cambridge University Press, 1906). Z. Kopal's *Widening Horizons* (Stanmore Press, London, 1971) gives a more recent and shorter account of several chapters of this early history. Of subsequent works, a delightful summary of the state of the subject in the mid-1930s, written in non-technical language, can be found in H. N. Russell's little book on *The Solar System and Its Origin* (Macmillan, New York, 1935).

The classics of the early 1950s were without doubt the books entitled *The Planets* (Yale University Press, 1952) by H. C. Urey, and *Origin of the Solar System* (Oxford University Press, 1954) by H. Alfvén. Alfvén and Urey have done more than anyone else in the past twenty years to place the modern solar-system research on a sound and imaginative physical basis.

Of contributions that have appeared in more recent years, the reader may still consult with profit the compendium on *The Solar System*, contributed by different authors, edited by G. P. Kuiper and B. M. Middlehurst, and published by the University of Chicago Press in 1961 ('Planets and Satellites'), and 1963 ('Moon, Meteorites, and Comets'). Although these volumes appeared at the beginning of the last decade, most articles included in them were written several years before and, therefore, reflect the state of the subject in the later 1950s.

Of books published in the last decade—when the entire subject of the solar-system studies was rejuvenated by the methods of space research—we should list S. W. McCuskey and V. M. Blanco's *Basic Physics of the Solar System* (Addison-Wesley Press, Reading, Mass., 1961); J. C. Brandt and P. W. Hodge's *Solar System Astrophysics* (McGraw Hill Publ. Co., New York, 1964); and W. M. Kaula's *Introduction to Planetary Physics* (John Wiley and Sons, Inc., New York, 1968) for a good account of basic methods underlying the current solar-system research. As far as our Moon is concerned, the reader seeking fuller information can consult Z. Kopal's treatise on *The Moon* (D. Reidel Publ. Co., Dordrecht; second edition,

1969) for a comprehensive account of the knowledge of our satellite in the pre-Apollo era.

On a more popular level, the reader may enjoy T. L. and L. W. Page's *The Origin of the Solar System* (Macmillan, New York, 1966) or F. L. Whipple's *Earth, Moon and Planets* (Harvard Univ. Press, third edition, 1969) which discuss the subject on a level similar to Russell's in 1935.

Such is the current rate of progress in solar-system exploration that an appreciable part of the material presented in this volume has been acquired since 1969 and cannot, therefore, be found in any of the books of earlier vintage referred to above.

Glossary

areographic Of the planet Mars (Ares was the Greek god of war).

albedo A measure of the fraction of light which is scattered from a surface.

aphelion The point of a planet's orbit farthest from the Sun.

apogee The point in the orbit of the Moon at which it is farthest from the Earth.

astroblemes Scars on the surface of a planet or satellite produced by the impact of meteorites, comets, or asteroids.

astronomical unit A unit of length equal to the mean distance between the Sun and the Earth (149 597 892 $\pm$ 5 km).

Cytherean Of the planet Venus.

declination The celestial altitude of a star or planet in equatorial coordinates—i.e. the angle by which it lies above or below the plane of the terrestrial equator.

direct rotation The rotation of a planet is direct if it is in the same sense (e.g. clockwise) as the planet revolves.

eccentricity *See* orbital.

ecliptic The orbital plane of the Earth around the Sun.

ephemeris Predicted position of a celestial body, contained in a table or almanac.

equation of state A relation between pressure and density (or temperature) in a given physical system.

gibbous A phase of an illuminated sphere between 'full' and the first or last quarter.

half-life The time taken for half of a sample of radioactive material to decay.

inferior conjunction Two planets are at inferior conjunction when they are at closest approach to one another.

invariable plane The plane which is perpendicular to the total momentum vector of the solar system as a whole; its inclination to the ecliptic is now close to $1°43'$.

isostasy The condition of equilibrium of surface pressure—thus on the Earth mountain ranges and their underlying crusts are less dense than the surrounding rocks.

light year The distance travelled by light in the course of one year. Since the velocity of light is 299 793 km s^{-1} and the length of the year is $3{\cdot}15\,569 \times 10^7$ s, one light year is $9{\cdot}46 \times 10^{12}$ km or 0·3169 parsecs.

limb of a planet The edge of a planet as seen from the Earth.

magnitude A measure of brightness of an object on a logarithmic scale. If Star A is of 2nd magnitude and Star B is of 3rd magnitude, then Star A is 2·514 times brighter than Star B. The absolute magnitude of a star is equal to its apparent brightness as seen from a standard distance of 10 parsecs.

meridian of a planet for the terrestrial observer The plane containing the terrestrial observer and the rotational axis of the planet.

nodes The points of intersection of the orbital plane of a planet or satellite with the ecliptic.

occultation The disappearance of a smaller body behind a larger one.

opposition Two planets are in opposition when they and the Sun lie on the same line (the two planets being on the same side). The outer of the two planets is then in *opposition* to the inner, while the inner is in *inferior conjunction* with the outer.

orbital eccentricity The deviation of an orbit from a circle as given by $(l - s)/(l + s)$ where l and s are the long and short diameters of the orbit.

orbital period The time taken for a planet or a satellite to complete one orbit round the central body.

parsec A unit of distance at which the mean radius of the Earth's orbit would subtend one second of arc (206 265 astronomical units or 3·262 light years).

perigee The point in the orbit of the Moon at which it is nearest to the Earth.

perihelion The point in a planet's orbit at which it is nearest to the Sun.

prograde Forward-moving.

quadrature A satellite or a planet is at quadrature with respect to the Earth if the Sun and the respective celestial body appear to us to be separated by 90° from each other in the sky.

regolith The mantle of loose rock-particles on the surface of a planet or satellite.

right-ascension The celestial longitude of a star or planet in equatorial coordinates.

secular Occurring over a very long period of time.

semi-major axis Equal to the mean radius of an eccentric orbit.

sidereal day The time taken by a planet to make one complete revolution about its axis relative to distant stars rather than to the Sun.

solar wind The stream of elementary particles which evaporates into space from the Sun.

steradian The unit of solid angle subtended at the centre by a segment of area r^2 of the surface of a sphere of radius r.

superior conjunction Two planets are at superior conjunction at the time of their greatest separation—i.e. when they are on opposite sides of the Sun from each other.

synodic A synodic time-interval is the time that elapses between successive identical phases of a planet or satellite.

terminator The dividing line between the light and dark part of the apparent disc of a planet or satellite.

Index